Space Physics

Space Physics

R. Stephen White

Professor of Physics
University of California, Riverside

GORDON AND BREACH, Science Publishers

New York · London · Paris

Editorial office for Great Britain:
Gordon and Breach, Science Publishers Ltd
12 *Bloomsbury Way*
*London W.C.*1

Editorial office for France:
Gordon & Breach
7–9 *rue Emile Dubois*
Paris 14e

Distributed in Canada by:
The Ryerson Press
299 *Queen Street West*
Toronto 2B, *Ontario*

To FREDA, my wife

Preface

In any rapidly growing field, there is a great need to correlate a large body of data from many disciplines and to unify many diverse ideas. This is especially true in Space Physics where the total volume of information has exploded in the last 10 years. Much unexpected data have been obtained since the invention of the rockets that boost satellites into space—data that have lead to many unpredicted discoveries.

An attempt is made in this book to condense and unify the large amount of space information and to summarize and criticize the currently accepted theories. It is hoped that the book will be useful as a textbook to the student who needs a broad background in the various subjects of Space Physics. It is hoped that it will be useful as a reference work to the research scientist who finds it difficult to keep up with all aspects of Space Physics, particularly outside his special field of research. And that it will be of use to the engineer, by describing the experiments that need the sophisticated space equipment.

In order for the book to be useful to more readers with different backgrounds, the mathematics is intentionally kept to a minimum. Experimental facts, historical discoveries and important ideas are emphasized.

Not all scientists will agree on the material that should be included in Space Physics. This book includes chapters on the Earth's Radiation Belts, Atmosphere, Ionosphere, Magnetic Cavity, the Sun and Interplanetary Space. The radiations from the sun and the interactions of these radiations with the earth are described. The lack of discussions of Mars, Venus and Jupiter and the Moon, about which we now have considerable data, is a serious omission. For this, I apologize and must request the reader to obtain the information from other sources, until the additional chapters can be added.

Since the book evaluates new data and new ideas in a fast growing field, there is the risk that the material will become obsolete before the book is published. Most of the ideas and concepts discussed are reasonably firm and should not change too rapidly in the next few years. But the reader is warned that in the future, as experiments on deep space probes explore the planets and move closer to the sun, we will find new discoveries that are beyond our present imagination—new discoveries that will challenge the present ideas of this fascinating new field of Space Physics.

Contents

Acknowledgements

The author wishes to thank the following for permission to reproduce figures: Academic Press, Inc. (Fig. 6.27); the American Association for the Advancement of Science (Figs. 1.9, 2.1, 3.5, 5.16, 6.1, 6.2, 6.17); the American Geophysical Union (Figs. 1.3, 1.8, 1.10, 1.13, 2.3, 2.18, 3.6, 3.11–16, 3.19, 3.20, 3.22, 3.23, 4.22, 4.25, 4.27–30, 5.3–15, 5.17–19, 5.21, 5.22, 5.24, 5.25, 5.30, 5.31, 6.4, 6.5, 6.8–14, 6.18–20, 6.22, 6.26); the American Institute of Aeronautics and Astronautics (Fig. 6.25); the American Institute of Physics (Figs. 5.20, 5.26, 6.21, 6.23); Annual Reviews, Inc. (Figs. 4.5, 4.11b, 4.23, 4.24); Cambridge University Press (Fig. 4.31); Centre National de la Recherche Scientifique (Figs. 2.4, 2.5, 3.7); the Commonwealth Scientific and Industrial Research Organisation (Fig. 4.18); W. H. Freeman and Co. (Figs. 1.1, 5.1, 6.15, 6.16); Holt, Rinehart and Winston, Inc. (Fig. 4.4); The Johns Hopkins Press (Fig. 6.7); McGraw-Hill Book Co. (Figs. 4.6, 4.16); Macmillan (Journals) Ltd (Figs. 1.15, 3.2, 3.8); National Aeronautics and Space Administration (Figs. 4.3, 4.19, 4.20, 4.26); the National Research Council of Canada (Fig. 6.24); North-Holland Publishing Co. (Figs. 1.16, 2.6–11, 3.18, 4.17); Observatoire de Paris (Figs. 4.9a, 4.9b); *Physics Today* (Figs. 1.2, 1.4, 1.11, 6.6); D. Reidel Publishing Co. (Figs. 5.23, 5.28, 5.29); the Royal Astronomical Society (Fig. 4.11a); The Royal Society (Figs. 3.3, 3.9, 3.17a); Sacramento Peak Observatory (Fig. 4.8); Solar Observatory, Aerospace Corporation (Figs. 4.9c, 4.13, 4.15); Space Research Council, Slough (Fig. 3.10); University of Chicago Press (Figs. 4.14, 4.32, 5.27); University of Minnesota (Fig. 4.12); Uppsala University Astronomical Observatory (Fig. 4.10c); Wadsworth Publishing Co., Inc. (Fig. 4.2); Yerkes Observatory (Figs. 4.10a, 4.10b).

CHAPTER 1

The Earth's Radiation Belts

1.1 INTRODUCTION

Ten years of space experimentation have elapsed since James A. Van Allen, George H. Ludwig, Carl E. McIlwain, and E. C. Ray[1] discovered the earth's radiation belts with a Geiger counter on the Explorer 1 satellite. Almost every flight since has offered the opportunity of

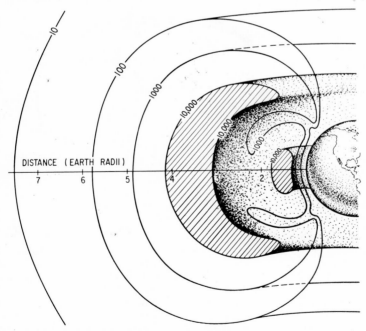

DISTANCE (EARTH RADII)

Figure 1.1. The inner and outer radiation belts taken from Van Allen,[3] 1959. The inner and outer belts are indicated by the slanted lines. Contours of count rates of 1,000, 100, and 10 are also indicated. The distances from the center of the earth in earth radii are labeled.

finding a different particle, a new energy distribution or an unusual space or time variation.

At first there were only two belts, an inner and an outer belt. The inner belt was centered at an altitude of 3,000 km. The outer belt, with a maximum at about 25,000 km, was added with the discovery of a second large radiation zone by a Geiger counter on the moon probe Pioneer 3.[2] These belts were beautifully displayed in a review article by Van Allen[3] in 1959 and are reproduced here in Fig. 1.1.

The shapes of the inner and the outer electron belts, in three dimensions, are sketched in Fig. 1.2 with the three types of charged particle motion superimposed. The particle circles the magnetic field line with

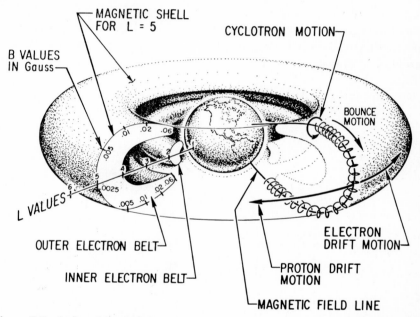

Figure 1.2. A drawing of the two electron radiation belts taken from White,[73] 1966. They appear horseshoe-shaped in the cutaway. The cyclotron motion about the magnetic field line, the bounce motion along the line, and the drift motion around the earth are shown. Protons drift westward, electrons drift eastward. A magnetic L shell is formed by rotating a magnetic field line about the earth's magnetic dipole axis. The deviations of the true magnetic field from a dipole field change the shape slightly. L is the distance to the shell at the equator in earth radii. Each particle always mirrors at its same value of B. The particle's motion is thus described by the L shell on which it moves and the B value at which it mirrors. The magnetic shell for $L = 5$ is labeled. B values in gauss on this shell are noted. L values on a radial line are indicated.

the familiar cyclotron motion, bounces back and forth along the magnetic field line and drifts around the earth on a magnetic L shell; the electrons drift east and the protons west. L is the distance to the magnetic shell at the equator, usually given in earth radii. Each L shell is formed by rotating a magnetic field line around the earth's magnetic dipole axis. Slight changes from this shape arise because of the deviations of the true magnetic field from a dipole field. Since the particle always mirrors at the same value of the magnetic field B as it drifts around on its L shell, its motion can be described by the L shell on which it moves and the B value at which it mirrors.

1.2 HIGH ENERGY PROTONS

The Geiger counter is an excellent exploratory detector and has been used with great success by Van Allen and co-workers. But it was not able to identify the particles in the radiation belts.

Early in 1959, Stanley C. Freden and R. Stephen White[4] had the opportunity of flying small stacks of nuclear emulsions on Air Force missile nose cones. These were recovered down range from Cape Kennedy after reaching an altitude of 1,200 km. With emulsions it was possible to distinguish among protons, alpha-particles and electrons and even deuterons, tritons and He^3 particles.

The first successful recovery of a Thor nose cone in April 1959 gave excellent results. The emulsions were loaded with proton tracks. The shielding around the emulsions stopped protons of less than 75 MeV or electrons of less than 12 MeV. No electrons were seen. Subsequent flights in May 1959 and October 1960[5-7] pushed the proton energy distribution down to 10 MeV. The resulting proton energy distribution is shown in Fig. 1.3. No electrons above 5 MeV were observed. Deuterons and tritons were only 1 per cent of the protons. No alpha-particles were detected. This was not surprising since the minimum detectable energy for alpha-particles was very high, 300 MeV.

The explanation[8] for the trapped protons first came from S. Fred Singer and independently from S. N. Vernov and colleagues. They recognized that the cosmic ray protons collide with atmospheric nuclei and that the product neutrons stream back into the magnetosphere and decay into protons, electrons, and neutrinos. The charged electrons and protons spiral around the magnetic field lines and move around the earth trapped on magnetic shells. The protons lose energy by ioniza-

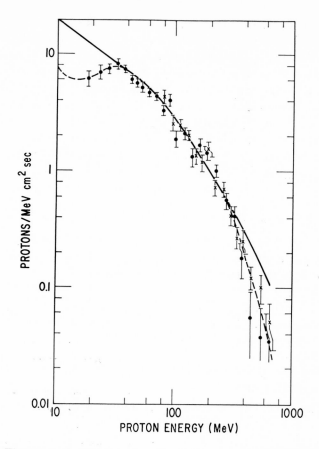

Figure 1.3. The trapped proton energy distribution as measured by Freden and White,[6] 1962. The shape of this distribution was well fit by cosmic ray albedo neutron decay injection and atmospheric ionization and nuclear collision losses. Recent calculations[9] show that this theory will give enough trapped protons to explain the data only if the ratio of neutron albedo flux to the atmospheric density is a factor of 50 larger than is currently estimated.

tion and excitation of atmospheric atoms, and an equilibrium distribution is then maintained between the cosmic ray albedo neutron decay source, CRAND, and the atmospheric sink. Singer predicted the shape of the proton energy distribution. Later another loss mechanism, nuclear interactions,[5] was added to the theory. This is significant for protons above 75 MeV. The CRAND injection into the high energy belt is shown in Fig. 1.4.

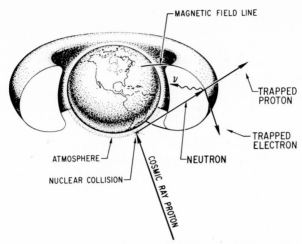

Figure 1.4. The high energy proton belts taken from White,[73] 1966. Actually the proton belts are displaced for different energies. 75 MeV protons peak at $L = 1.5$ but lower energy protons peak farther out. The cosmic ray albedo neutron decay injection, CRAND, is shown. The cosmic ray proton makes a nuclear collision in the atmosphere and emits a neutron. The neutron decays into a proton, electron, and neutrino. The electron and proton are trapped in the earth's magnetic field.

At equilibrium, the number of protons/cm³-sec-MeV injected must equal the number lost. Then

$$\chi \frac{j_n}{\gamma v \tau} = \frac{d}{dE}\left(j_p \frac{dE}{dx} \right) + j_p \sigma \frac{\rho A_0}{A} \qquad (1.1)$$

where j_n is the albedo neutron flux at the position of injection, v is the neutron velocity (also the proton velocity since it carries away almost all of the neutron momentum).

$$\gamma = 1 \Big/ \sqrt{1 - \left(\frac{v}{c}\right)^2},$$

τ is the neutron lifetime of 1.0×10^3 sec, and χ is the injection coefficient, the probability that the proton from the neutron decay makes a sufficiently large angle with the magnetic field direction that the proton is initially trapped. The dE/dx is the ionization energy loss per unit path length and is proportional to the mean atmospheric density ρ, averaged over the proton's trajectory. The j_p is the trapped proton flux, and σ is the nuclear collision cross section. The A is the mean atomic weight and the A_0 is Avogadro's number. If other losses are significant they must be added to the right side of Eq. (1.1.).

Consider the simplest case, trapped protons of less than 100 MeV with ionization loss only. Take power laws in energy for j_n, γv, dE/dx and j_p. Then the equilibrium proton flux is

$$j_p = \frac{C\chi j_n(E)E^{1.28}}{\rho}.$$

(1.2)

We see that j_p is proportional to the injection coefficient χ, and to j_n, and is inversely proportional to the mean atmospheric density averaged over the particle trajectory ρ. C is a constant. Before the theoretical expected equilibrium flux can be determined, χ, ρ, and j_n must be evaluated.

The early calculations took χ equal to one, a big overestimate. Later computations[9] that simulate the ejection of neutrons from the earth have been made with a computer program. Various neutron latitude and zenith angle distributions were used. χ varies somewhat with these distributions and with position in space but on the average is about 0.1 for the CRAND source.

Trapped protons can also be injected from solar proton albedo neutron decays, SPAND. The solar protons arise from solar flares. The computed χ for SPAND injection is even lower than for CRAND because neutrons from the poles are less likely to inject trapped protons than neutrons from the equator. The uncertainties in the χ's, for the cases considered, are much less than 50 per cent since the calculations are strictly geometric.

The ρ is obtained by averaging the atmospheric density over a proton drift period around the earth.[10] A model atmosphere and a magnetic field representation are required for the computation. At a few hundred kilometers near the lower edge of the trapped radiation belt the atmospheric densities are known from satellite drag measurements to better than a factor of two. However, the densities change rapidly with altitude. The exponential decreasing distance is approximately 50 km, so the proton trajectories through this atmosphere must be precisely known. The currently used 48 and 512 term representations of the earth's magnetic field give differences as large as a factor of 10 in the mean densities.[10] Above 1,000 km altitude, the magnetic field differences are less important but the atmospheric densities are not as well known. Local atmospheric anomalies or changes with latitude and longitude have not been well investigated. The uncertainty in j_p due to ρ in some regions of space can be as large as a factor of 10.

Singer[8] estimated the albedo neutron flux from the proton reaction fragments of cosmic ray proton interactions in nuclear emulsions. Wilmot H. Hess,[11] on the other hand, used the neutron flux below 10 MeV, measured in the atmosphere at airplane altitudes[12] with BF_3 counters, and connected this to the cosmic ray spectrum at 1 BeV. The input neutron spectrum between 10 and 100 MeV is necessary for comparison between experiment and theory but is currently the poorest known part of the theory.[9] A factor of 10 uncertainty in the albedo neutron flux is realistic.

Late in 1959, on Explorer, 6, C. Y. Fan, Peter Meyer, and J. Arol Simpson[13] found that the protons decreased rapidly at distances greater than 4,000 km above the surface of the earth. This decrease could not be explained by the CRAND source, since CRAND injection should only vary slowly in that region of space. Singer[14] suggested that this was due to the breakdown of the first adiabatic invariant, the magnetic moment

$$\mu = \frac{E_\perp}{B}. \tag{1.3}$$

$E_\perp$ is the proton energy associated with the velocity perpendicular to the magnetic field B. At large distances from the earth, time or space variations in B can be as large as the steady magnetic field. If these variations occur in times or regions of space which are small compared to the proton cyclotron period or radius, μ invariance is violated. The protons are then driven down into the denser atmosphere where they are quickly lost. Dragt[15] suggested that hydromagnetic waves are responsible for the breakdown. He calculated the losses and found these reasonable to account for the low proton fluxes at great distances from the earth. Consequently, an additional term representing diffusion in pitch angle, the angle between the proton direction and the magnetic field caused by electromagnetic wave interactions should be added to the equilibrium equation for the trapped proton flux.

John E. Naugle and D. A. Kniffen[16] obtained energy distributions of protons with nuclear emulsions at a number of altitudes on the probe NERV in September 1960. These distributions were of particular interest because they showed that protons with energies of 10 MeV increased rapidly with altitude. These protons were first observed to increase at the altitude and space location where neutrons could decay after straight line travel from the earth's polar cap. Since the earth's polar cap is bombarded by protons from the sun at times of solar flares and the solar flare proton energy distribution is much softer than the cosmic ray primary spectrum, the SPAND injection was considered a prime candidate for the source of the low energy protons. Allen M. Lenchek[17] computed the flux and energy distribution to be expected and concluded that SPAND was indeed a likely source. There should be no SPAND injected protons for $L < 1.5$ and there should be no SPAND injected protons at higher L values at the equator. Neither of these predictions seem to be verified by later measurements with scintillators and solid state detectors.[18-20] The abrupt increase of the flux at $L = 1.5$ suggested by the emulsion measurements is not found. In addition, the low energy protons are more abundant at the equator than at higher B values close to the earth. The discrepancies between the experiments and the theory and between the emulsion and the counter measurements have not yet been resolved.

Although the energy distributions measured with nuclear emulsions are still the most detailed, they have one serious drawback. Nuclear emulsions must be recovered before the data can be analyzed. For that reason they are not useful for extended spatial distributions. The spatial measurements have been obtained with counters that read-out directly into telemetry over a receiving station or onto a tape recorder which is played into telemetry when the satellite passes over a receiving station.

The constraints forced upon space experiments are usually quite severe. The experiments must be lightweight, typically from 0.5 to 5 Kg and must draw low power, typically 1 watt. Telemetry is seldom as much as desired. No multi-ton magnets or large shielding blocks considered essential around accelerators are permitted. There is neither space nor power for huge banks of scalers, amplifiers, and power supplies. After launch of the experiments no discriminators can be tweeked or circuits repaired. It is not possible to stop, re-design the experiment and try again at a later date. For these reasons the identification of the radiation belt particles, and measurements of their fluxes and energies have proceeded slower than measurements on the ground.

A typical spin stabilized spacecraft is shown in Fig. 1.5. One detector is drawn mounted on the side. As the spacecraft spins the detector

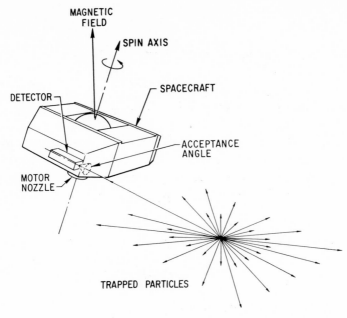

Figure 1.5. A typical spacecraft for radiation belt experiments. A particle detector is mounted perpendicular to the spin axis, which is shown at a small angle to the local magnetic field. Trapped particles are concentrated in a disk perpendicular to the magnetic field. As the spacecraft spins, the particle flux measured by the detector goes through two maxima per revolution.

points into and out of the disc of trapped particles that are perpendicular to the magnetic field direction. The spin modulates the particle flux signal measured by the detector. If the acceptance angle is small the modulation can be as great as a factor of 100.

A block diagram of a spacecraft data system is drawn in Fig. 1.6. Experiment data go to commutators which feed two tracks of a tape recorder. A 1×60 commutator samples 60 analog data points every second and the $1/2 \times 60$ commutator samples 60 points every two seconds. A subcommutator samples spacecraft status information like temperatures and voltages and feeds to one point of the 1×60 commutator. The stored tape-recorded data is read out on command in FM–FM channels A and C over a receiving station. Real time data is also read out through channels 9 and 11. Power for the experiments and the spacecraft is furnished by solar cells.

Experiments are sometimes mounted as shown in the photograph of

DATA SYSTEM

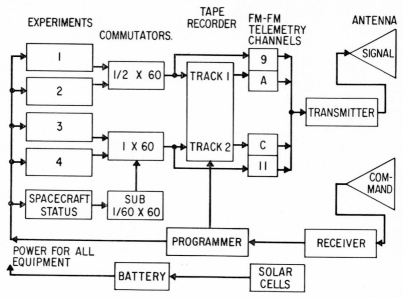

Figure 1.6. A typical experiment data system. Data from the experiments go to $\frac{1}{2} \times 60$ and 1×60 commutators which feed two tracks of a tape recorder. The tape recorder is read out by command into FM/FM telemetry channels A and C over a receiving station. Real time data is also obtained in channels 9 and 11. Power for the experiments and for the spacecraft is furnished by solar cells which are mounted on panels and fastened to the spacecraft frame.

the satellite 1964–45A in Fig. 1.7. The detectors are all perpendicular to the spacecraft spin axis which is perpendicular to the plane of the paper. An electron spectrometer, a proton spectrometer and five omni-directional detectors for electrons and protons are clearly visible. The rocket motor nozzle points upward. The spin-up jets are indicated. Solar cells are mounted on flat planes and attached to the spacecraft frame.

In addition to the difficulties of instrumentation are the difficulties that the trapped radiation belts are not constant in time. One nice simplification to the data handling has been added by McIlwain.[21] The three space coordinates, latitude, longitude and altitude, are replaced by just two, B and L. L is a function of B and of the second adiabatic invariant J, defined as:[22]

$$J = \oint p_{\parallel} \, ds, \tag{1.4}$$

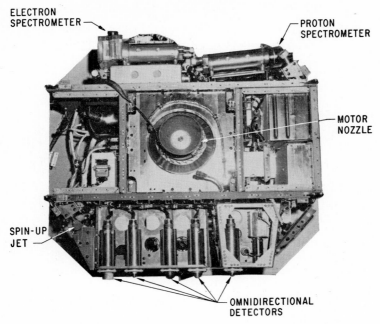

Figure 1.7. Photograph of the detectors mounted on the spacecraft 1964–45A. The experiment detector apertures are mounted perpendicular to the satellite spin axis, which is perpendicular to the plane of the paper. Shown are an electron spectrometer, a proton spectrometer, and five omnidirectional detectors for electrons and protons. The rocket motor nozzle and the spin-up jets are indicated.

where $p_{\parallel}$ is the momentum parallel to the magnetic field and ds is measured along the line. The line integral is taken over one complete bounce motion from one mirror point to its conjugate and back again.

The data is often presented on B–L plots like the one of Fig. 1.8, which shows the Explorer 4 protons of McIlwain.[21] Contours of constant counting rate are plotted as solid lines. These contours start from the equator parallel to lines of constant altitude (not shown) but then turn downward. The lines of constant altitude would continue upward toward the right side of the graph. Lines of constant latitude would have curvatures similar to the equator and to its right. An equivalent representation in latitude λ, and distance from the center of the earth R, was previously shown in Fig. 1.1.

Plastic scintillators[18] with fixed thresholds show that the proton energy distributions vary greatly with position in space. The two peaks

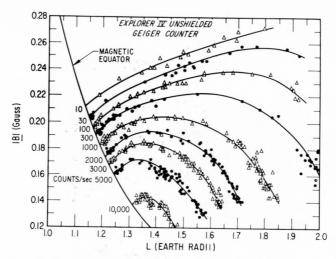

Figure 1.8. A B–L plot of contours of constant counting rate taken from McIlwain,[21] 1961. Data in three dimensions (latitude, longitude, and distance) are reduced to two dimensions (the mirroring magnetic field, B, and the magnetic shell, L, on which the particles move). The contours of constant counting rate start from the equator parallel to the lines of constant altitude (not shown) but then turn downward. The lines of constant altitude would continue upward toward the right. The lines of constant latitude would be similar to the equator and to its right.

found in the curve of flux as a function of L for protons with energies of 40 to 80 MeV are demonstrated in Fig. 1.9. The outer peak was found at $L = 2.2$ and the inner peak as before at $L = 1.5$. Neither CRAND nor SPAND explains the double peaks. If higher energy protons had been plotted the $L = 1.5$ peak would have been closer in. And the peak for lower energy protons would have been farther out.

Additional energy distributions and fluxes as a function of B and L have been measured with plastic and solid state detectors[19, 20] and with emulsions.[23,24] These distributions are not explained either. In fact, recent CRAND calculations[9] of the fluxes are too low by a factor of 50 to explain the trapped protons at low altitudes and SPAND is an additional factor of 10 lower. A possible explanation is that the ratio of the currently accepted albedo neutron flux to the mean atmospheric density encountered by the trapped protons used in the calculations is a factor of 50 too low. A measurement of the earth's albedo neutron energy distribution in the energy range of 10 to 100 MeV is badly needed.

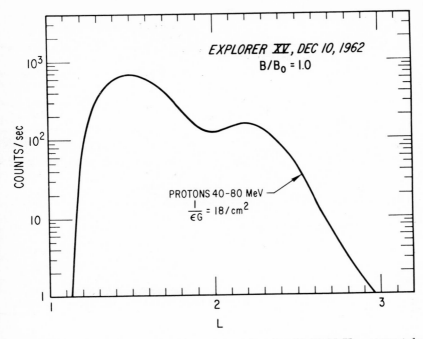

Figure 1.9. A plot of the count rate at the equator for 40–80 MeV protons taken from McIlwain,[18] 1963. This shows two maxima of the 40–80 MeV protons. The first maximum is at $L = 1.5$ and the second is at $L = 2.2$. The data are taken at the equator, $B/B_0 = 1.0$. The geometrical factor, $1/\epsilon G$, is 18/cm².

Small variations in time of the high energy proton belts for $L < 2.5$ were observed and shown to be adiabatic by McIlwain.[25] Because of geomagnetic disturbances the proton intensities decrease slightly and then return to their pre-storm values in a few days. An example of this adiabatic behavior is the storm on April 18, 1965[25] shown in Fig. 1.10. The proton intensity decrease in per cent per γ (10^{-5} gauss) was 0.05 per cent/γ at $L = 1.5$ and 0.2 per cent/γ at $L = 2.1$ at the exact time of the decrease in the magnetic field at the earth's surface. The explanation of this adiabatic behavior is that a ring current is formed at distances greater than about three earth radii from the earth. The trapped particles undergo betatron acceleration because of the ring current, and carry out a slight radial movement. This causes the initial loss of trapped particles. The ring current then decays slowly and the trapped particles return to their initial positions.

An apparent non-adiabatic change[23] in the high energy proton belt at

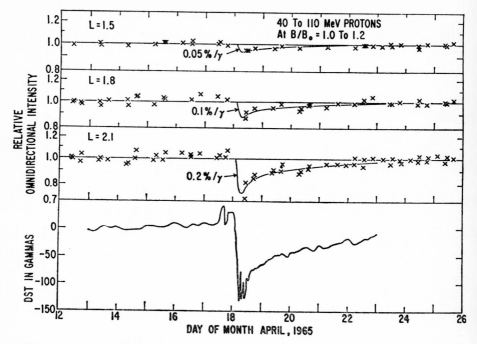

Figure 1.10. The adiabatic changes in the high energy proton belt at the time of a magnetic storm taken from McIlwain,[25] 1966. The decrease coincident with the decrease in the ring current magnetic field and return to normal is shown for $L = 1.5$, 1.8, and 2.1.

low altitudes took place as a result of the Starfish bomb explosion on July 9, 1962. The number of 55 MeV protons at an altitude of 300 km in the South Atlantic increased abruptly by a factor of 4 and then slowly returned to their pre-bomb values with a decay constant of about a year. This is explained by the big change in the magnetic field which accompanied the explosion. The pitch angles of some of the trapped protons were decreased so that they mirrored at altitudes closer to the earth. This increased the flux of protons at lower altitudes. Gradually the excess protons at low altitudes were lost by collisions with atmospheric atoms and the high energy proton belt returned to normal.

It is likely that another source is responsible for protons with energies less than 10 MeV. This leads us to the low energy proton distributions and the inward diffusion source.

1.3 LOW ENERGY PROTONS

The low energy proton belts surround the earth in multilayers like concentric onion skins. The lowest energy protons are on the outside and the higher energy ones are closer in. Three of the proton belts are shown in Fig. 1.11 at $L = 4.5$, 3.5, and 2.5. The belts are shown as separated, but they run continuously from one energy to the next.

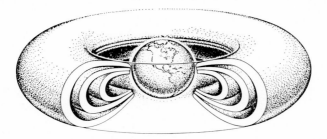

Figure 1.11. A drawing of the low energy proton radiation belts which peak at $L = 4.5$, 3.5, and 2.5 taken from White,[73] 1966. The belts for each energy overlap and run smoothly into each other so that there are no gaps. The lowest energy protons peak at high L values and higher energy protons peak closer in. These belts probably originate from diffusion inward from the solar wind across magnetic field lines.

Intense fluxes of these low energy protons were discovered by Leo R. Davis and James M. Williamson[26] on Explorer 12 which was launched in August 1961. They found 6×10^7 protons/cm²-sec-ster on the equator at $L = 3.5$ with $E > 0.1$ MeV. Integral energy distributions measured from 0.1 to 1.6 MeV, became harder at lower L values. They obtained additional measurements from Explorers 14 and 15. The energy distribution from 1 to 10 MeV was measured by Sam J. Bame and his coworkers[27] on a rocket launched in October 1960, which went along a magnetic field line at $L = 2$.

Paul J. Kellogg[28] had pointed out in 1959 that particles diffusing inward at the equator while conserving the first adiabatic invariant should increase their energies E according to

$$EL^3 = \text{const.} \tag{1.5}$$

Equation (1.5) is obtained by substituting the expression for a magnetic dipole field, $B = \mathcal{M}/L^3$, into Eq. (1.3). The $\mathcal{M}$ is the magnetic moment of the earth.

In 1960 Eugene N. Parker[29] invented a mechanism for obtaining the diffusion. The third adiabatic invariant, which is the total magnetic flux through the drift orbit, is violated. Sudden impulses in the magnetic field, which come in a time short compared to the drift period of the protons, cause betatron acceleration in the drift orbit. The drift period is inversely proportional to E and L and is about 15 minutes. The acceleration is about 10 keV for a large storm. The orbits are pushed inward for some protons and outward for others. The relaxation back to the original condition occurs adiabatically and this causes a random walk in L. This diffusion is illustrated in Fig. 1.12. The

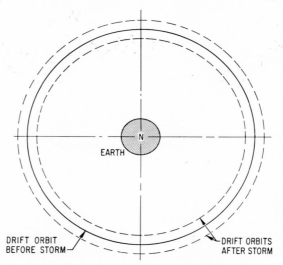

Figure 1.12. The diffusion of particles by magnetic storms suggested by Parker,[29] 1960. Protons initially on the drift orbit of the solid line diffuse inward and outward to the dashed lines because of a magnetic storm.

solid line shows the initial drift orbit. Protons on that orbit are diffused inward and outward by a magnetic storm to orbits indicated by the dashed lines. Parker found a diffusion constant that increased outward as L^{10}. These ideas were applied [30] to calculate the constant in the exponential of the proton integral energy distribution. It was found to increase toward lower L values in agreement with the measured value.

M. Paul Nakada and G. D. Mead[31] used a Fokker–Planck diffusion equation to obtain the theoretical proton spatial and energy distributions along the equator. They included diffusion coefficients[32] for the mean value of the displacement in R and R^2, proportional to L^9 and L^{10}, respectively, an exponential integral energy distribution at the edge of the magnetosphere, and Coulomb energy and charge exchange losses.

The theory can be compared to the data of Mihalov and White[33] who measured the differential spatial and energy distributions of protons at 12 energies between 0.2 and 6 MeV with a CsI spectrometer on the satellite 1964–45A. They found that the lowest measured energy—0.19 MeV—has its maximum value at $L = 4.5$, whereas the highest energy of 2.8 MeV has its maximum at $L = 2.4$. These maxima appeared to be independent of B, so they could be applied to the equator. The values of EL^3 indicate that the first invariant is conserved up to an energy of 0.75 MeV but not at higher energies. Off the equator $EL^3 \sin^2 \alpha_0(L)$ should be constant if both the first and second adiabatic invariants are conserved. The pitch angle at the

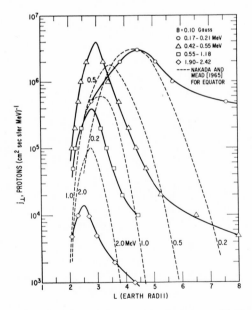

Figure 1.13. A plot of the low energy proton flux perpendicular to the magnetic field line versus the L magnetic shell taken from Mihalov and White,[33] 1966. The solid lines are experimental data taken at $B = 0.10$ gauss and at the energies indicated on the face of the graph. The dashed lines are the theoretical predictions of Nakada and Mead,[31] 1965, for the equator and for values of the energy in MeV shown next to the dashed lines. The theory is normalized to the data at 0.2 MeV at $L = 4.4$.

equator $\alpha_0(L)$ is evaluated along a drift trajectory. Since the experimental $EL^3 \sin^2 \alpha_0(L)$ deviates even more from a constant than EL^3 it is also concluded that the first and second invariants are not conserved for energies above 0.75 MeV.

The theoretical spatial and energy distributions[31] were derived for positions on the equator only, and were not extended to the B values of the Mihalov–White[33] experiment. Nevertheless, the general features of the theory do not appear to depend strongly on this fact; so the theory is compared to the experimental data at $B = 0.10$ gauss. These general features are in good agreement with the experimental ones. The detailed comparisons of the spatial distributions of Fig. 1.13 and the energy distributions of Fig. 1.14, however, show some important

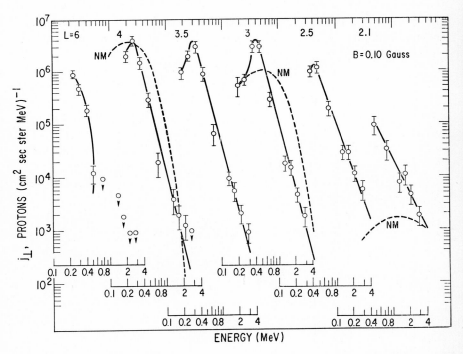

Figure 1.14. Energy distributions of low energy protons taken from Mihalov and White,[33] 1966. The flux of protons perpendicular to the magnetic field line is plotted versus the proton energy at $B = 0.10$ gauss for $L = 6$, 4, 3.5, 3, 2.5, and 2.1. Solid lines are drawn through the data. The downward pointing arrow indicates an upper limit. The dashed curves are the theoretical predictions of Nakada and Mead,[31] 1965, normalized to the experimental data at $L = 4$ for an energy of 0.25 MeV.

differences. The theoretical curves, which were normalized to the experimental ones at $L = 5$, $E = 0.20$ MeV and $B = 0.10$ gauss, peak at higher L values than the experimental distributions and fall off less rapidly with increasing energy.

The experimental energy distributions[33] of Fig. 1.14 have narrow peaks. Indeed, the true energy peak widths must be very narrow because the observed widths are comparable to the energy resolutions. If these protons are due to diffusion inward, a peaked proton source of 15 keV at $L = 10$ at the edge of the magnetosphere is required. In the theoretical energy distributions the turnover at low energies is caused by the proton losses only, and not by the source since an exponential energy source at $L = 10$ at the edge of the magnetosphere was assumed.

At low L values, the theoretical distributions are too low at low energies; that is, at $L = 2.0$ and $E = 0.40$ MeV the theoretical value is a factor of 100 lower than the experimental value. The agreement between theory and experiment would be improved by using a larger diffusion rate.

At low L values diffusion is so slow and the proton losses so great that the fluxes fall extremely rapidly between $L = 3.25$ and $L = 2.35$. For $L < 2.0$, it seems very difficult to account for the protons by diffusion inward from the magnetospheric boundary.

It is interesting that the straight-line power law fits to the energy distributions of Fig. 1.14 are also in reasonable agreement with the data at 55 MeV.[19] This agreement is over an energy range of a factor of 100 and a flux range of 2×10^7.

A comparison of experiments up to 1966 indicates that the low energy proton belts have been rather constant in time since 1962. In a 6-month period from August 1964 to February 1965,[34] protons of 0.2 MeV decreased slightly, protons of 1 MeV remained constant, and those of 3 MeV increased a little. This probably indicated a slight redistribution of the protons among the magnetic shells during that time. Coincident with the previously mentioned negative bay of the magnetic storm on April 18, 1965, low energy protons[35] increased by a factor of 10, then returned to normal with an apparent lifetime of a few hours. Other small storms were previously observed to give small changes in the low energy protons.

Finally, we should emphasize that the diffusion theories have not calculated the absolute number of protons injected into the earth's magnetic field. Consequently, a prediction of the absolute number of

B

trapped low energy protons has not been made. Until the injection mechanism is known and absolute equilibrium fluxes can be calculated, the explanation for these protons is inadequate.

1.4 ELECTRONS

Today there are two great electron belts—the inner belt and the outer belt. These are shown in Figs. 1.1 and 1.2. The inner belt consists mostly of the fission electron remains of Starfish, the United States high altitude nuclear burst which was exploded in July 1962. This belt is now peaked at $L = 1.35$. These electrons do not fluctuate, only slowly decrease with time.

The outer belt is peaked at $L = 5$ and is separated from the inner belt by a large slot. It is made up of natural electrons of lower energies. It varies considerably with time and is correlated with changes in the magnetic field. How the electrons got there is still an unanswered question.

Extensive surveys have been carried out by Van Allen and his co-workers. They have measured the spatial distributions of particles in great detail on Explorers 1, 4, 7, 12 and 14.[1,2,36−39] They found a flux of 10^8 electrons/cm²-sec[36] with energies greater than 40 kev in the outer belt in September 1961. Since that time there have been large time variations but the flux is usually within a factor of 10 of 10^7 electrons/cm²-sec. The average value has been rather constant over the last eight years.

K. I. Gringauz and coworkers[40] discovered fluxes of 10^7 electrons/cm²-sec with energies greater than 200 eV with a Faraday Cup on Lunik 2 at 50,000 to 80,000 km from the earth in a backward direction of 45 deg to the sun–earth line. They also observed the low energy electrons on Lunik 1 which had a trajectory of about 75 deg to the sun–earth direction. The authors considered the electrons as a third or outermost belt. These electrons are now thought to be part of the magnetosphere tail or transition region (see Chapter 6).

Also on Lunik 2,[41] approximately 5×10^5 electrons/cm²-sec were found with energies between 1 and 2 MeV. These measurements suggested that the previous Geiger counter measurements[2,13] were really due to penetrating electrons and not to bremsstrahlung as had been supposed.

No reliable measurements were made of the electron fluxes at the equator at the maximum of the inner radiation belt before July 1962. However, the natural electrons of the inner radiation belt were identified and their energies were measured by Atlas pods at altitudes of 1,500 km in 1959.[42] The differential energy spectrum measured at $L \simeq 1.3$, $B \simeq 0.25$ decreased by a factor of 20 from 100 keV to 450 keV.

The electrons of the outer belt were identified and their energies measured on Javelin rockets.[43] Between 50 and 700 keV, an exponential energy distribution was found which drops by $1/e$ in 60 keV. This was considerably steeper than the one in the inner radiation belt.

Calculations were made by Hess and his colleagues[44] using CRAND injection and atmospheric energy loss and scattering. They concluded that one could explain the electrons on the basis of CRAND injection if the electron lifetimes at the equator were longer than about ten years. They also computed the theoretical energy distributions but found them flatter than the measured ones.[45] Relatively more electrons were found experimentally at low altitudes than predicted by the neutron decay theory.[46]

The outer belt electron fluxes were found experimentally to increase and then return to normal in times as short as a month.[37] Measurements by B. J. O'Brien[47] of large fluxes of electrons with small pitch angles indicate that such large fluxes of electrons are continuously lost into the atmosphere that the lifetimes have to be short. Therefore, CRAND injection seemed impossible for the outer radiation belt. And the slot between the two belts was not explained.

Before more sophisticated measurements could clarify the injection and loss mechanisms, Starfish injected enough electrons into the inner radiation belt to entirely mask the natural electrons. Starfish exploded at an altitude of 400 km on July 9, 1962 over Johnson Island in the Pacific. The yield was 2 MT and 2×10^8 fission electrons/cm²-sec were injected at the maximum of the inner radiation belt at $L = 1.3$. Shortly thereafter, the Russians followed with three nuclear injections on October 22, 28, and November 1 at L values of 1.88, 1.81, and 1.77.[48,49]

These were not the first nuclear explosions in space. Following the suggestion of Christofilos[50] in an unpublished memorandum in 1957 that many geophysical effects could be observed, three nuclear explosions, called Argus, were carried out by the Advanced Research Projects Agency. The 1 to 2 kiloton bursts occurred on August 27, 30, and September 6, 1958 at 480 km over the South Atlantic.[51] Experiments[51] on Explorer 4 found that the drift in latitude of the Argus III shell at $L = 2.2$ was < 0.03 deg latitude/day or < 1 km/day. They found that the count rate fell off as $1/t$ for the first 10 days. The fission electrons were also observed with experiments[52] on sounding rockets and a similar decay was found.

Since the electrons were injected into the radiation belts at known times it was possible to follow the electron fluxes as a function of time and to measure the electron lifetimes. Lifetimes for the Russian electrons were measured to be only a few days.[18,37,48,49,53] The lifetimes for Starfish electrons initially were just as short but quickly increased to about one year.[54-56] These lifetimes are summarized in Table 1.1. The lifetime[56] as a function of L is given in Fig. 1.15. Martin Walt[57] calculated the expected lifetimes using a Fokker–Planck diffusion equation and losses by scattering and ionization in the atmosphere. These agreed very well with the experimental decay curves[56] from a few hours up to 50 days for $0.18 < B < 0.22$ gauss for

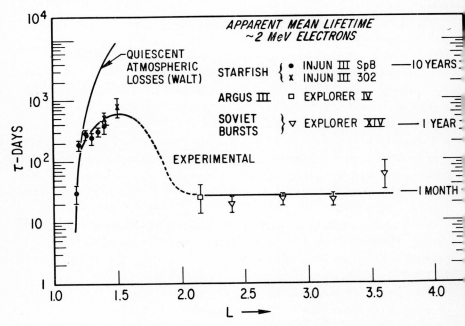

Figure 1.15. A plot of the lifetime versus the L magnetic shell taken from Van Allen,[56] 1964. The data points are from Injun 3 and Explorers 4 and 14. The theoretical lifetime, computed on the basis of atmospheric losses by Walt,[57] 1964, is included.

	L (Earth radii)	B (Gauss)	Date	Electron energy (MeV)	Mean lifetime (days)	Authors
Starfish July 9, 1962	2.0	0.04–0.08	July 20, 1962	> 0.2	15	Brown and Gabbe[54]
	2.2	0.05–0.10			4	
	2.5	0.06–0.14			5	
	3.0	0.07–0.17			4	
	3.5	0.10–0.20			14	
	1.5	0.094	Dec. 10, 1962	> 0.5	∞	McIlwain[18]
	2.0	0.40			50	
	1.5	0.094		> 5.0	∞	
	2.0	0.40			50	
	1.2	0.185	1963	> 0.5	140	McIlwain—Reported in Walt[57]
	1.3	0.146			270	
	1.4	0.119			270	
	1.5	0.094			270	
	1.20	0.185–0.205	Dec. 22, 1963	> 1.2	120 ±12 130 ±20	Bostrom and Williams[55] Van Allen[56]
	1.23	0.170–0.205			165 ±50 190 ±30	
	1.30	0.160–0.230			235 ±20 320 ±70	
	1.40	0.165–0.210			390 ±40 500 ±100	
	1.50	0.175–0.215			460 ±50 590 ±300	
	1.60	0.180–0.225			360 ±50	
Russian Oct. 22, 1962 Oct. 28, 1962	2.8	0.017–0.038	Oct. 28, 1962– Feb. 14, 1963	> 1.6	30	Frank, Van Allen, Hills[37]
	2.80–3.10	0.330–0.355	Oct. 22, 1962 Oct. 28, 1962 Nov. 1, 1962 / Nov. 10–30, 1962 Dec. 10–30, 1962	> 3.9	20 50	Burrows, McDiarmid[48]*
	1.77	0.252	Nov. 6–10, 1962	0.35 bremsstrahlung	7^{+4}_{-2}	Mihalov, Mozer, and White[49]

* Burrows and McDiarmid fit their decay data to a power series in time $I = I_0\, t^{-\eta}$. Over a period of two months following three injections $\eta \simeq 1.3$ at $L = 2.05$, $B = 0.240$; $L = 2.20$, $B = 0.250$, $L = 2.55$, $B = 0.260$; and $L = 2.95$, $B = 0.270$. The values appearing in the table are exponential approximations at the times indicated.

$L < 1.3$, which indicated that the atmosphere controls the lifetimes there. But for $L > 1.3$ the calculated lifetimes were longer than the experimentally measured values.

At a particular position in B-L space, electrons which were injected initially are augmented with those from lower B values (higher altitudes) by scattering. Electrons are lost by the same process. The atmospheric density decreases rapidly as B decreases (the altitude increases). Transient equilibrium is established between the electrons at high and low altitudes. The problem is similar to radioactive decay where the parent–daughter relationship is given by

$$j_d = \frac{j_\pi}{\tau_\pi} \tau_d. \qquad (1.6)$$

j is the flux of electrons/cm²-sec, τ is the mean lifetime in sec, π signifies the parent and d the daughter. The j_d decreases at the same rate as $j\pi$ which falls off exponentially with a lifetime of τ_π. From Table 1.1 it can be seen that transient equilibrium was reached rather rapidly in most parts of space. The maximum lifetime is measured at the equator at $L = 1.5$ and is about two years. Higher and lower L values give shorter lifetimes. If the atmosphere and electron trajectories in the earth's magnetic field are not greatly different from those used by Walt,[57] another non-atmospheric loss mechanism must become important for electrons for $L \gtrsim 1.3$.

Detailed electron energy distributions were measured at 5 energies in a 180 deg focusing magnetic spectrometer.[53] The measurements were made on the satellite 1962 βK (STARAD) which was launched October 26, 1962. They showed that the Starfish electrons had fission energy distributions. The electron spectra of the Russian injections were similar. For higher L, the energy distributions were steeper. Typical electron energy distributions[53] measured at low L values are given in Fig. 1.16.

At low altitudes William L. Imhof and R. V. Smith[58] obtained an unexpected narrow peak[59] at 1.3 MeV, $L = 1.15$ and $B = 0.217$, which decayed away with a time constant of one day. The time constant is just that expected from atmospheric losses. This decay may be explained by time oscillations in the earth's magnetic field.[60] These oscillations are in resonance with the drift period of specific energy electrons, which diffuse to lower L values. Nearly monoenergetic electrons then appear at positions in space where none previously existed.

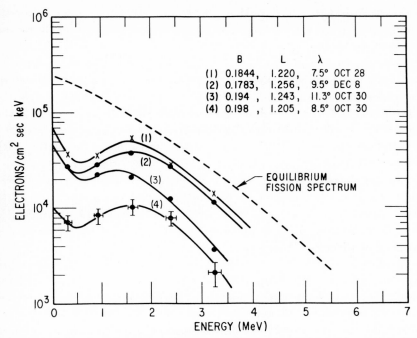

Figure 1.16. The electron energy distribution taken from West, Mann, and Bloom,[53] 1965, for measurements in 1962. The solid lines are drawn through the data points at B, L, and magnetic latitudes listed on the face of the graph. An equilibrium fission spectrum is shown as the dashed line. This data is for the peak of the Starfish belt.

After the Starfish and the Russian injections the slot was temporarily filled but rapidly emptied again to separate the inner from the outer radiation belt. Walter L. Brown and J. D. Gabbe[54] reported a time constant of 5 days for $2.2 \leqslant L \leqslant 3.0$. There must be large electron losses in this region of space. If the loss were atmospheric only, a large peak in the atmospheric density contours between $L = 2$ and $L = 3$ would be required. Such a peak has not been identified. The only latitude effect appearing in present atmospheric models is the sun's diurnal variation.

In the fall of 1962 on low altitude polar satellites, George A. Paulikas and Stanley C. Freden[61] observed semi-trapped fluxes of 10^3 electrons/cm²-sec with energies > 0.9 MeV at $L = 1.2$ and $L = 2.0$. These electrons were lost when they dipped deep into the atmosphere in their drift trajectories over the South Atlantic. They probably leaked out of the Starfish belt.   Perhaps these electrons were driven down magnetic field lines by resonant electromagnetic wave interactions that decreased the electron pitch angles.[62]

Large fluctuations in electron intensities as a function of time and of position have been observed in the outer radiation belt.[38]   Electrons have been observed by Louis A. Frank[63] to diffuse inward at the rate of $0.4(R_e/\text{day})$ at $L = 4.7$ decreasing to 0.03 (R_e/day) at $L = 3.4$ (R_e is the radius of the earth).   In this L region the diffusion rate was measured to be proportional to L^8.   Flux changes have been correlated with magnetic storms.[64,65]

Are the electrons lost from the magnetosphere into the atmosphere? Or do they change position in the magnetosphere and then return to their original positions?   Are the electrons accelerated to higher energies only to relax back to their initial energies?   What is the electron loss rate out of the magnetosphere?

Where the electron lifetime is very long, only a few of the electrons leak out of the magnetosphere and a weak source such as CRAND could be responsible for injection.   There, significant numbers of observed electrons with energies higher than the neutron decay limit of 0.8 MeV could result from diffusion inward while conserving the first adiabatic invariant, satisfying Eq. (1.5). Then the observed fluctuations would not be true losses, but only temporary changes in the equilibrium distribution.

On the other hand, where the lifetime is short, the source must be strong and the only seriously considered contender is the solar wind. The solar wind must furnish the electrons which diffuse inward because of magnetic storms.   The energy increases with decreasing L by Eq. (1.5). Although diffusion appears to work for low energy protons (Section 1.3), the situation is not so obvious for the electrons which do not obey Eq. (1.5).[66]   If diffusion does occur a source of 100 keV electrons must exist at $L = 10$ at the edge of the magnetosphere.   If the solar wind is responsible, the electrons need to be accelerated from the solar wind energy of 1 eV to 100 keV at $L = 10$.   Thus, the problem is still not solved, only shifted to finding how and where solar wind electrons are accelerated.   An acceptable alternative is to find a source of 100 keV electrons which can work their way into the magnetosphere. To date no such source has been identified.

The detailed energy distributions of Mihalov and White[66] suggest that the Starfish injected fission electrons are still dominant at high energies for $L < 1.7$ and a natural soft enegy distribution appears above the fission electrons.

The slot is a feature common to all energies, at least above 40 keV. The slot was present in the early data and was formed again rapidly after Starfish. The electron flux increases to a maximum in the outer belt at $L = 5$ and decreases at higher L values. The energy distribution gradually becomes softer as L is increased.

The latest measurements (1968) show for the first time the diffusion of electrons inward to low L values in the inner radiation belt. K. A. Pfitzer and J. R. Winckler[67] on the satellite OGO–III observed the diffusion of electrons into the inner zone to $L = 1.3$ following the solar disturbance on September 2, 1966. The initial wave of electrons penetrated to $L = 2.2$ within a day and to $L = 1.4$ in about 40 days. Following this diffusion a new and stable inner zone of electrons with energies of less than 690 keV was formed. No evidence was found for the diffusion of electrons with energies greater than 690 keV to L values of less than 2.0. The new inner belt created by the injection followed a Starfish-like decay after equilibrium was reached.

Because of the 1 year lifetime for these electrons, one or two injections of this magnitude per year could supply the inner zone with low energy electrons of 50 to 690 keV.

Rapid pitch angle distribution changes were recently reported (1968) by Donald Williams, J. F. Arens and L. J. Lanzerotti[68] from experiments on the satellites 1963 38C and Explorer 26. Equilibrium along a magnetic field line was attained for electrons with energies of greater than 300 keV in 0.1 day. The mechanisms responsible could be cyclotron-resonance scattering by whistler-mode disturbances or bounce-resonance scattering by disturbances having electric or magnetic field components parallel to the local field[69] (see Chapt. 3).

The 1 MeV electrons take significantly longer to attain equilibrium than the 300 keV electrons. Therefore, the pitch-angle mechanisms are not as effective for the 1 MeV electrons. Since the cyclotron-resonance frequency is about 1000 c/sec at $L = 3$ and 200 c/sec at $L = 5$ for these electrons, the lack of pitch angle diffusion may mean that the whistler electromagnetic noise decreases significantly below 1000 c/sec.

1.5 MAGNETIC CAVITY EFFECT ON L

The previous use of the magnetic shell number L and the mirror magnetic field B to label electrons and protons is nearly exact out to an L value of 5. However, at larger L values, the magnetic field departs significantly from a dipole field and is no longer symmetric with longitude because of the long tail of the earth's magnetic cavity. See Chapt. 6 for a detailed description of the earth's magnetic field. Because of this non-symmetry in the earth's magnetic field, particles starting on the same magnetic field line on the sun side of the earth at noon but at different mirror point altitudes populate different magnetic shells in the earth's shadow at midnight. Likewise particles starting on a particular magnetic field line in the earth's shadow at midnight separate onto many different shells on the sun side at noon.

Juan Roederer[70] studied the motion of energetic particles in a model magnetosphere that included the earth's magnetic tail. He arrived at the following conclusions:

(1) Shell splitting in the outer magnetosphere becomes important beyond $5R_E$; dipole-type descriptions of the radiation belt become invalid.

(2) Equatorial pitch angles tend to align along field lines on the night side of the magnetosphere and perpendicularly to the field on the day side.

(3) There are regions in the magnetosphere where only pseudo-trapped particles can mirror, i.e. particles that will leave the magnetosphere before completing a 180 deg drift.

(4) Longitudinal drift velocities depart considerably from the dipole values beyond $5R_e$, and they can be as much as 2–3 times greater on the night side than on the day side. Thus a given particle spends 2–3 times more time in the day side than in the night side.

(5) The action of a pitch-angle scattering mechanism will lead to a radial diffusion of particles. The loss mechanism will be greatly enhanced by scattering of mirror points into the pseudo-trapping regions.

(6) After recovery from a prototype magnetic storm, particles that were in the day side during the sudden commencement will have higher energies, their shells having moved radially inward. Particles caught in the night side will have moved outward, with their energies decreased.

(7) The repeated action of magnetic storms will result in a net inward diffusion of particles, with a net increase of their energy.

1.6 SUMMARY

It is useful to summarize the great variety and volume of data that has been accumulated over the last ten years. In one such study, James I. Vette has[71] compiled a composite trapped inner zone radiation environment. Composite B–L and R–λ flux maps for electrons with energies > 0.5 MeV and for protons with energies greater than 4, 15, 34, and 50 MeV have been prepared. These will be updated from time to time. Flux maps at other energies and at higher L values are currently under preparation. For the man in space programs it is necessary to know the accumulated radiation dosages for specific flight missions. This radiation dosage has been computed for a number of typical orbits through the radiation belts.

An attempt has been made to emphasize the limitations of the experiments and the theories. These limitations can be further delineated by reading the comprehensive review articles,[72] in particular the review[73] which contains much of the information presented here. Much has been learned about the radiation belts since their discovery in 1958. But out knowledge is still inadequate to answer the most basic questions. What are the sources and what are the losses of particles in the radiation belts? Perhaps we will find the answers during the next maximum of the solar cycle—during its peak in 1970–71.

REFERENCES

1. Van Allen, J. A., Ludwig, G. H., Ray, E. C. and McIlwain, C. E., "Observations of high-intensity radiation by satellites 1958 alpha and gamma," *Jet Propulsion* **28**, 588–592 (1958).
2. Van Allen, J. A. and Frank, L. A., "Radiation around the earth to a radial distance of 107,400 km," *Nature* **183**, 430–434 (1959).
3. Van Allen, J. A., "Radiation belts around the earth," *Scientific American* **200**, No. 3, 39–47 (1959).
4. Freden, S. C. and White, R. S., "Protons in the earth's magnetic field," *Phys. Rev. Letters* **3**, 9–10 (1959).
5. Freden, S. C. and White, R. S., "Particle fluxes in the inner radiation belt," *J. Geophys. Res.* **65**, 1377–1383 (1960).
6. Freden, S. C. and White, R. S., "Trapped proton and cosmic ray albedo neutron fluxes," *J. Geophys. Res.* **67**, 25–29 (1962).
7. Heckman, H. H. and Armstrong, A. H., "Energy spectrum of geomagnetically trapped protons," *J. Geophys. Res.* **67**, 1255–1262 (1962).

8. Singer, S. F., "Radiation belt and trapped cosmic ray albedo, " *Phys. Rev. Letters* **1**, 181–183 (1958); Vernov, S. N., Grigorov, N. L., Ivanenko, I. D., Lebedinskii, A. I., Murzin, V. W. and Chudakov, A. E., "Possible mechanism of production of terrestrial corpuscular radiation under the action of cosmic rays," *Soviet Phys. Dokl.* **4**, 154–157 (1959).

9. Dragt, A. J., Austin, M. M. and White, R. S., "Cosmic ray and solar proton albedo neutron decay injection," *J. Geophys. Res.* **71**, 1293–1304 (1966).

10. Cornwall, J. M., Sims, A. R. and White, R. S., "Atmospheric density experienced by radiation belt protons," *J. Geophys. Res.* **70**, 3099–3111 (1965).

11. Hess, W. N., "Van Allen belt protons from cosmic ray neutron leakage," *Phys. Rev. Letters* **3**, 11–13 (1959).

12. Hess, W. N., Patterson, H. W., Wallace, R. and Chupp, E. L., "Cosmic ray neutron energy distribution," *Phys. Rev.* **116**, 445–475 (1959).

13. Fan, C. Y., Meyer, P. and Simpson, J. A., "Dynamics and structure of the outer radiation belt," *J. Geophys. Res.* **66**, 2607–2640 (1961).

14. Singer, S. F., "Cause of the minimum in the earth's radiation belt," *Phys. Rev. Letters* **3**, 188–190 (1959).

15. Dragt, A. J., "Effect of hydromagnetic waves on the lifetime of Van Allen radiation protons," *J. Geophys. Res.* **66**, 1641–1649 (1961).

16. Naugle, J. E. and Kniffen, D. A., "Variations of the proton energy spectrum with position in the inner radiation belt," *J. Geophys. Res.* **68**, 4065–4078 (1963).

17. Lenchek, A. M., "On the anomalous component of low energy geomagnetically trapped protons," *J. Geophys. Res.* **67**, 2145–2157 (1962).

18. McIlwain, C. E., "The radiation belts, natural and artificial," *Science* **142**, 355–361 (1963).

19. Fillius, R. W. and McIlwain, C. E., "The anomalous energy spectrum of protons in the earth's radiation belt," *Phys. Rev. Letters* **12**, 609–611 (1964).

20. Freden, S. C., Blake, J. B. and Paulikas, G. A., "Spatial variation of the inner zone trapped proton spectrum," *J. Geophys. Res.* **70**, 3111–3117 (1965).

21. McIlwain, C. E., "Coordinates for mapping the distribution of magnetically trapped particles," *J. Geophys. Res.* **66**, 3681–3691 (1961).

22. Northrup, T. G. and Teller, E., "Stability of the adiabatic motion of charged particles in the earth's field," *Phys. Rev.* **117**, 215–225 (1960).

23. Filz, R. C. and Holeman, E., "Time and altitude dependence of 55 MeV trapped protons, August 1961 to June 1964," *J. Geophys. Res.* **70**, 5807–5822 (1965).

24. Heckman, H. H. and Nakano, G. H., "Direct observations of mirroring protons in the south atlantic anomaly," *Space Research V*, p. 329–342, North-Holland Publishing Co., Amsterdam, 1964.

25. McIlwain, C. E., "Ring current effects on trapped particles," *J. Geophys. Res.* **71**, 3623–3628 (1966).

26. Davis, L. R. and Williamson, J. M., "Low energy trapped protons," *Space Research III*, p. 365–376, North-Holland Publishing Co., Amsterdam, 1963.

27. Bame, S. J., Conner, J. P., Hill, H. H. and Holly, F. E., "Protons in the outer zone of the radiation belt," *J. Geophys. Res.* **68**, 55–63 (1963).

28. Kellogg, P. J., "Van Allen radiation of solar origin," *Nature* **183**, 1295–1297 (1959).

29. Parker, E. N., Geomagnetic fluctuations and the form of the outer zone of the Van Allen radiation belt," *J. Geophys. Res.* **65**, 3117–3130 (1960).

30. Dungey, J. W., Hess, W. N. and Nakada, M. P., "Theoretical studies of

protons in the outer radiation belt," *Space Research IV*, p. 399–403, North-Holland Publishing Co., Amsterdam, 1965.

31. Nakada, M. P. and Mead, G. D., "Diffusion of protons in the outer radiation belt," *J. Geophys. Res.* **70**, 4777–4791 (1965).

32. Davis, L., Jr. and Chang, D.B., "On the effect of geomagnetic fluctuations on trapped particles," *J. Geophys. Res.* **67**, 2169–2179 (1962).

33. Mihalov. J. D. and White, R. S., "The low energy proton radiation belts," *J. Geophys. Res.* **71**, 2207–2216 (1966).

34. White, R. S., "The time dependence of the low energy proton belts," *J. Geophys. Res.* **72**, 943–950 (1967).

35. Davis, L. R. and Williamson, J. M., "Outer zone protons," p. 215–230, *Radiation Trapped in the Earth's Magnetic Field*, Proceedings of the NATO Advanced Study Institute, Bergen, Norway, August 16-September 3, 1965, McCormac, H. M. (ed.), D. Reidel Publishing Co., Dordrecht, Holland, 1966.

36. O'Brien, B. J., Van Allen, J. A., Laughlin, C. D. and Frank, L. A., "Absolute electron intensities in the heart of the earth's outer radiation zone," *J. Geophys. Res.* **67**, 397–404 (1962).

37. Frank, L. A., Van Allen, J. A. and Hills, H. K., "A study of charged particles in the earth's outer radiation zone with Explorer 14," *J. Geophys. Res.*, **69**, 2171–2191 (1964).

38. Freeman, J. W., "The morphology of the electron distribution in the outer radiation zone and near the magnetospheric boundary as observed by Explorer 12," *J. Geophys. Res.* **69**, 1691–1723 (1964).

39. Frank, L. A., "A survey of electrons $E > 40$ keV beyond 5 earth radii with Explorer 14," *J. Geophys. Res.* **70**, 1593–1626 (1965).

40. Gringauz, K. I., Bezrukikh, V. V., Ozerov, V. D. and Kybchinskii, R. E., "The study of ionized gas, energetic electrons, and corpuscular emission of the sun with the aid of charged particle three-electron trap installed on the second soviet space rocket," *Dokl. Akad. Nauk SSSR* **131** (6), 1301 (1960); Gringauz, K. I., Kurt, V. G., Moroz, V. I. and Shklovskiy, I. S., "Ionized gas and fast electrons in the vicinity of the earth," *Dokl. Akad. Nauk SSSR* **132** (5), 1062 (1960).

41. Vernov, S. N., Chudakov, A. E., Valsulov, P. V., Logachev, Yu. I. and Nikolayev, A. G. "Radiation measurements during the flight of the second soviet space rocket," *Space Research I*, p. 845–851, North-Holland Publishing Co., Amsterdam, 1960; originally published in *Dokl. Akad. Nauk SSSR* **130**, 517–520 (1960).

42. Holly, F. E., Allen, L. and Johnson, R. G., "Radiation measurements to 1500 kilometers altitude at equatorial latitudes," *J. Geophys. Res.* **66**, 1627–1639 (1961).

43. Cladis, J. B., Chase, L. F., Jr., Imhof, W. L. and Knecht, D. J., "Energy spectrum and angular distributions of electrons trapped in the geomagnetic fields," *J. Geophys. Res.* **66**, 2297–2312 (1961).

44. Hess, W. N., "The radiation belt produced by neutrons leaking out of the atmosphere of the earth," *J. Geophys. Res.* **65**, 3107–3115 (1960); Hess, W. N. and Killeen, J., "Spatial distribution of electrons from neutron decay in the outer radiation belt," *J. Geophys. Res.* **66**, 3671–3680 (1961); Hess, W. N., Canfield, E. H. and Lingenfelter, R. E., "Cosmic ray neutron demography," *J. Geophys. Res.* **66**, 665–679 (1961).

45. Hess, W. N. and Poirier, J. A., "Energy spectrum of electrons in the outer radiation belt," *J. Geophys. Res.* **67**, 1699–1709 (1962).

46. Hess, W. N., Killeen, J., Fan, C. Y., Meyer, P. and Simpson, J. A., "The observed outer-belt electron distribution and the neutron decay hypothesis," *J. Geophys. Res.* **66**, 2313–2314 (1961).

47. O'Brien, B. J., "Direct observations of dumping of electrons at 1000 kilometer altitude and high latitudes," *J. Geophys. Res.* **67**, 1227–1235 (1962).

48. Burrows, J. R. and McDiarmid, I. B., "A study of electrons artificially injected into the geomagnetic field in October, 1962," *Can. J. Phys.* **42**, 1529–1549 (1964).

49. Mihalov, J. D., Mozer, F. S. and White, R. S., "Artificially injected electrons at low altitudes," *J. Geophys. Res.* **69**, 4003–4013 (1964).

50. Christofilos, N. C., "The Argus experiment," *J. Geophys. Res.* **64**, 869–875 (1959).

51. Van Allen, J. A., McIlwain, C. E. and Ludwig, G. H., "Satellite observations of electrons artificially injected into the geomagnetic field," *J. Geophys. Res.* **64**, 877–891 (1959).

52. Allen, L., Jr., Beavers, II, J. L., Whitaker, W. A., Welch, J. A., Jr. and Walton, R. B., "Project Jason measurements of trapped electrons from a nuclear device by sounding rockets," *J. Geophys. Res.* **64**, 893–908 (1959).

53. West, H. I., Jr., Mann, L. G. and Bloom, S. D., "Some electron spectra in the radiation belts in the fall of 1962," *Space Research V*, p. 423–445, North-Holland Publishing Co., Amsterdam, 1965.

54. Brown, W. L. and Gabbe, J. D., "The electron distribution in the earth's radiation belts during July 1962 as measured by Telstar," *J. Geophys. Res.* **68**, 607–618 (1963).

55. Bostrom, C. O. and Williams, D. J., "Time decay of the artificial radiation belt," *J. Geophys. Res.* **70**, 240–243 (1965).

56. Van Allen, J. A., "Lifetimes of geomagnetically trapped electrons of several MeV energy," *Nature* **203**, 1006–1007 (1964).

57. Walt, M., "The effects of atmospheric collisions on geomagnetically trapped electrons," *J. Geophys. Res.* **69**, 3947—3958 (1964).

58. Imhof, W. L. and Smith, R. V., "Energy spectrum of electrons at low altitudes," *J. Geophys. Res.* **70**, 2129–2135 (1965).

59. Imhof, W. L. and Smith, R. V., "Observation of nearly monoenergetic high energy electrons in the inner radiation belt," *Phys. Rev. Letters* **14**, 885–887 (1965).

60. Cladis, J. B., "Acceleration of geomagnetically trapped electrons by variations of ionospheric currents," *J. Geophys. Res.* **71**, 5019–5025 (1966).

61. Paulikas, G. A. and Freden, S. C., "Precipitation of energetic electrons into the atmosphere," *J. Geophys. Res.* **69**, 1239–1249 (1964).

62. Cornwall, J. M., "Scattering of energetic trapped electrons by very low frequency waves," *J. Geophys. Res.* **69**, 1251–1259 (1964); Dungey, J. W., "Loss of Van Allen electrons due to whistlers," *Planetary Space Sci.* **11**, 591–596 (1963).

63. Frank, L. A.: "Inward radial diffusion of electrons of greater than 1.6 million electron volts in the outer radiation zone," *J. Geophys. Res.* **70**, 3533–3540 (1965).

64. Pizzella, G., McIlwain, C. E. and Van Allen, J. A., "The time variations of intensity in the earth's inner radiation zone, October 1959 through December 1960," *J. Geophys. Res.* **67**, 1235–1253 (1962)

65. Forbush, S. E., Pizzella, G. and Venkatesan, D., "The morphology and temporal variations of the Van Allen radiation belt, October 1959 to December 1960," *J. Geophys. Res.* **67**, 3651–3668 (1962).

66. Mihalov, J. D. and White, R. S., "Some energetic electron spectra in the radiation belts," *J. Geophys. Res.* **71**, 2217–2226 (1966).
67. Pfitzer, K. A., and Winckler, J. R., "Experimental observation of a large addition to the electron inner radiation belt following a solar flare event," *J. Geophys. Res.* **73**, 5792–5798 (1968).
68. Williams, D. J., Arens, J. F., and Lanzerotti, L. J., "Observations of trapped electrons at low and high altitudes," *J. Geophys. Res.* **73**, 5673–5696 (1968).
69. Roberts, C. S., "Electron loss from the Van Allen zones due to pitch-angle scattering by electromagnetic disturbances," *Earth's Particles and Fields*, McCormac, B. M. (ed.), Reinhold Publishing Co., New York (1968).
70. Roederer, Juan G., "On the adiabatic motion of energetic particles in a model magnetosphere", *J. Geophys. Res.* **72**, 981–992 (1967).
71. Vette, J. I., *Models of the trapped radiation environment, Vol. I: Inner zone protons and electrons,*; Vette, J. I., Lucero, A. B. and Wright, J. A., *Models of the trapped radiation environment. Vol. II: Inner and outer zone electrons,* (National Aeronautics and Space Administration, Washington, D.C., 1966) Report No. NASA Sp-3024. This report is available upon request.
72. Van Allen, J. A., "Dynamics, composition and origin of the geomagnetically trapped radiation," *Trans. Intern. Astron. Union* **11***B*, 99–136 (1962); Hess, W. N., "Energetic particles in the inner Van Allen belt," *Space Sci. Rev.* **1**, 278–312 (1962); Farley, T. A.: "The growth of our knowledge of the earth's outer radiation belt," *Rev. Geophys.* **1**, 3–34 (1963); O'Brien, B. J., "Review of studies of trapped radiation with satellite-borne apparatus," *Space Sci. Rev.* **1**, 415–484 (1962–1963); Haerendel, G., "Protonen im Inneren Strahlungsgürtel," *Fortschritte der Physik* **12**, 251–346 (1964); Walt, M. and MacDonald, W., "The influence of the earth's atmosphere on geomagnetically trapped particles," *Rev. Geophys.* **2**, 543–577 (1964); LeGalley, D. P. and Rosen, A., Editors, *Space Physics* (John Wiley and Sons, Inc., New York, 1964); Hess, W. N., Mead, G. D. and Nakada, M. P., "Advances in particles and field research in the satellite era," *Rev. Geophys.* **3**, 521–570 (1965); McCormac, B. M., (ed.), *"Radiation trapped in the earth's magnetic field:* Proceedings of the NATO Advanced Study Institute, Bergen, Norway, August 16–September 3, 1965," D. Reidel Publishing Co., Holland (1966). Hess, Wilmot N., *The Radiation Belt and Magnetosphere*, Blaisdell Publishing Company, Waltham, Massachusetts (1968).
73. White, R. S., "The earth's radiation belts," *Physics Today* **19**, No. 10, 25–38 (1966).

CHAPTER 2

The Earth's Atmosphere

2.1 INTRODUCTION

Without an atmosphere man could not survive on earth. The atmosphere contains the oxygen that we breathe and the carbon dioxide for the plants that we eat. It furnishes protection against cosmic rays and holds in the heat of the sun. In space or on the moon man must bring along his atmosphere to live.

From the beginning of recorded time man has been interested in the weather. Meteorology dates back to Aristotle's "Meteorologica," 384–322 B.C., and to his pupil Theophrastus, who discussed the winds and the weather signs. But modern meteorology begins with instruments to record the properties of the weather, with the invention of the thermometer by Galileo in 1607 and the barometer by Torricelli in 1643.

The weather is the direct result of the heating and cooling of the lower atmosphere. This causes evaporation and condensation of atmospheric moisture and the differences in pressure which cause the atmospheric winds. The heating comes from the absorption of the sun's radiation, which is about 2 calories per min per cm^2 area perpendicular to the sun's rays. This input is about 5,000,000 horsepower per mi^2 (see, for example, Willet and Sanders[1]). About 34 per cent of the incident solar energy is reflected or scattered back into space and 66 per cent is absorbed by the earth and the atmosphere. Since the earth is at an average temperature of about 270° Kelvin, it reradiates energy entirely in the infrared instead of in the visible. The earth is much cooler than the 6,000° K of the photosphere of the sun. Most of the reradiated energy is absorbed by the water vapor in the atmosphere and the clouds and is retained. The trapping of the sun's radiation is called the Greenhouse Effect.

Since 1960 the weather has been observed from above by Tiros,

Essa and Nimbus satellites with television and infrared detectors. Excellent infrared measurements from Nimbus I[2] give the temperatures of the cloud tops, oceans, ice formations and the earth's surface. The heights of the tops of clouds can be calculated from their temperatures. And the dry, sandy ground with small heat capacity can be distinguished from the solid rocky surfaces with large heat capacity.

In Fig. 2.1 is a photograph of Hurricane Gladys taken from Nimbus I over the Atlantic during the night of September 17, 1964. A single

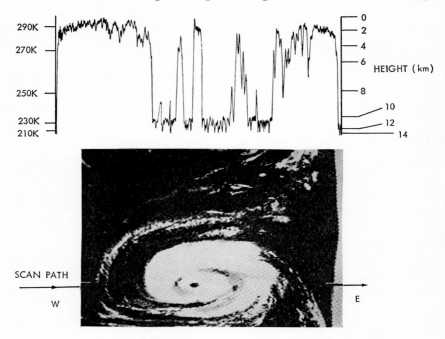

Figure 2.1. Photograph of Hurricane Gladys from the satellite Nimbus I over the Atlantic during the night of September 17, 1964. Taken from Nordberg,[2] 1965. A radiometer scan 5 km wide across the center of the hurricane is shown. The left vertical scale gives the blackbody temperature in degrees Kelvin and the right scale, the height obtained from these temperatures.

radiometer scan 5 km wide across the center of the hurricane is included. The left vertical scale gives the blackbody temperature in degrees Kelvin and the right scale, the heights obtained from these temperatures. The sea surface temperature under clear skies outside the storm was 300° K while the tops of the clouds inside the eye of the hurricane were at 290° K. Temperatures in the central parts of the spiral bands drop to 200° K corresponding to heights near 14 km.

The atmosphere is subdivided into regions whose boundaries are determined by a reversal in the temperature with height. In the troposphere the temperature decreases from 290° K on the earth to 220° K at the tropopause. In the stratosphere the temperature increases to 270° K then decreases to 180° K at the top of the mesopause. In the thermosphere the temperature increases rapidly and levels off at about 1500° K at a height of about 300 km during the middle of the solar cycle.

The weather—rainstorms, tornadoes, hurricanes—occurs in the troposphere whose upper boundary varies from 16 km at the equator to

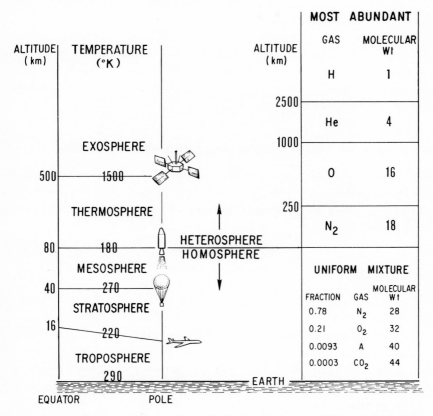

Figure 2.2. A diagram of the regions of the atmosphere and the most abundant gases. The left half shows the regions of the atmosphere, the right side, the most abundant gases and their molecular weights. Below 120 km there is a uniform mixture of the major gases, except for water, which varies with height.

8 km at the poles as sketched in Fig. 2.2. Jet planes often fly above the troposphere to avoid the weather. Balloons are sent aloft to maximum heights of about 40 km to collect meteorological data. Consequently, atmospheric measurements of temperature, pressure, atmospheric composition and wind were limited to about 40 km before the use of rockets following World War II. Below 80 km, the top of the mesosphere, the atmospheric composition, except for ozone and water vapor, is always the same, independent of height, and is called the homosphere. Above 80 km, collisions among molecules decrease. The earth acts like a giant gaseous mass separator. The reduced density and the force of gravity cause a mass separation, the heavy molecules settle to the bottom and the light ones rise to the top. The entire region above 120 km is sometimes called the heterosphere.

The region from 120 to about 450 km is often called the thermosphere because of the large increase in temperature there. Above 450 km the temperature is nearly constant and the atmospheric density is so low that collisions among neutral molecules are negligible. This region is called the exosphere.

2.2 PRESSURE, TEMPERATURE, AND DENSITY RELATIONS

The relations among temperature, pressure, and number density are obtained as follows. (See, for example, Francis S. Johnson.[3]) The differential pressure on a small volume of gas is given by the hydrostatic equation

$$dp = -\rho g \, dh, \tag{2.1}$$

where density ρ, is

$$\rho = \sum_i n_i m_i. \tag{2.2}$$

The h is the height above the surface of the earth, and g is the acceleration of gravity. The n_i are the number of molecules/cm³ of the molecular gas i with mass m_i. The "partial" pressure of each molecular gas species p_i, can be computed independently of the others.

The perfect gas law is

$$p_i = n_i kT, \tag{2.3}$$

where T is the absolute temperature in degrees Kelvin and k is the

Boltzmann constant, 1.38×10^{-16} erg/K°. Combining Eqs. (2.1) and (2.3) we have

$$\frac{dp_i}{p_i} = -\frac{m_i g \, dh}{kT} \tag{2.4}$$

and integrating

$$\frac{p_i}{p_{0_i}} = \exp\left(-\frac{m_i}{k}\int_0^h \frac{g}{T}\, dh\right) \tag{2.5}$$

where the 0 indicates the starting height. Since

$$n = \sum_i n_i \tag{2.6}$$

and the mean molecular weight is

$$M = \frac{\sum n_i m_i}{\sum n_i}. \tag{2.7}$$

The pressure, temperature, and number of molecules/cm³ are related by

$$\frac{p}{p_0} = \frac{nT}{n_0 T_0} = \frac{M_0}{M}\frac{\rho}{\rho_0}\frac{T}{T_0}. \tag{2.8}$$

If T, m, and g are independent of height,

$$\frac{p}{p_0} = e^{-mgh/kT} = e^{-h/H} \tag{2.9}$$

where the scale height H, is defined as

$$H = \frac{kT}{mg}. \tag{2.10}$$

In the case of helium or hydrogen where thermal diffusion may be important because of temperature gradients the number density is approximately[4]

$$n_i = n_{0_i}\left(\frac{T_0}{T}\right)^{1+q_i} \exp\left(-\frac{m_i}{k}\int_0^h g\,\frac{dh}{T}\right) \tag{2.11}$$

where q_i is the coefficient of thermal diffusion.

When the pressure varies horizontally a pressure gradient exists and the atmosphere moves to equalize the pressure. The equilibrium wind occurs when the coriolis force due to the rotation of the earth is equal to the pressure-gradient force. Then the wind velocity is

$$v = \frac{1}{2\rho\Omega\sin\lambda}\, \nabla_H p \, ,$$ (2.12)

where $\nabla_H p$ is the horizontal pressure gradient, Ω is the angular rotation velocity of the earth and λ is the latitude. Excluding tornadoes, the variation in pressure at the earth's surface is always less than or equal to 7 per cent. At higher altitudes the variations are greater; i.e., at 200 km the pressures vary by about a factor of 2. There the pressure-equalization flows greatly reduce the pressure variations that would otherwise exist.

2.3 EXPERIMENTAL MEASUREMENTS

Since 1946, after the development of rockets during World War II, it has been possible to measure atmospheric densities at heights previously unattainable. Prior to this time, composition, temperature and pressure information came from indirect aurora, nightglow, twilight glow and meteor observations. With rockets it was possible to fly ion gages, falling spheres, mass spectrometers and ultraviolet light absorption experiments to measure atmospheric densities. And since 1957 it has been possible to obtain the atmospheric density at the low point of the elliptical orbits of satellites by means of atmospheric drag. Barbara K. Ching and Robert A. Becker[5] have discussed and evaluated the density measurements up to the time of 1965.

2.3.1 Ionization gages

The first atmospheric density measurements at altitudes from 100 to 200 km were obtained in 1947 with Philips ionization gages.[6] Measurements were taken from 1947 to 1958 over White Sands, New Mexico at 32° N latitude and over Fort Churchill, Canada, at 59° N latitude. A general review of those measurements is given by H. E. LaGow, R. Horowitz and J. Ainsworth.[7] An ionization pressure gage measures the density of neutral particles inside its volume. The atmospheric density outside the gage is derived as a known function of the density inside. Two critical problems are always encountered, the method of connecting the detector to the atmosphere and the relative velocity between the detector and the atmosphere. When the gage faces ahead it sweeps out particles with velocities equal to the spacecraft velocity added to the gas thermal velocity. When it looks aft, the vehicle velocity subtracts from the gas thermal velocity. Consequently, in a spinning or tumbling spacecraft, the signal is modulated. Out-gassing of the pressure gage is often a nuisance and is the major limitation at high altitudes on rocket flights. Rocket propellant gas carried along with the rocket must be avoided. Recombination of gases on the walls of the gage is also usually a problem.

2.3.2 Dropping spheres

The first successful inflatable sphere experiment[8] was released from a rocket over the Elgin Gulf Test Range in Florida in 1961. The atmospheric density is obtained from an accurate measurement of the sphere's acceleration. The basic working equation[9] is

$$F_d = ma_d = \frac{1}{2}\rho v^2 A C_d \mathscr{F} \qquad (2.13)$$

where

F_d = drag force on the sphere,
m = mass of the sphere,
a_d = drag acceleration,
ρ = atmospheric density,
v = velocity of the sphere,
C_d = coefficient of drag,
A = cross-section area of the sphere, and
$\mathscr{F} \simeq 1$, a factor to correct for rotation of the atmosphere in satellite drag measurements.

Since m, A, and C_d are known, and a_d and v are measured, it is possible to find the atmospheric density ρ. The velocity and drag acceleration are found from ground tracking of the passive sphere, or with accelerometers attached to the sphere. Spheres from 18 to 270 cm in diameter made from aluminum and polyesters have been used during flight. The altitude resolution is ± 1 km. The technique is useful up to 135 km.

2.3.3 X-Ray and UV absorption

The absorption of solar X-rays[10] and extreme ultraviolet UV radiation[11] is used to obtain the atmospheric density of atmospheric constituents as a function of height. The intensity of the vertical radiation at the detector is found from

$$I = I_0 \exp\left(-\int_h^{h_\infty} \sigma n_i \, dh\right),$$ (2.14)

where:

I = intensity of radiation detected at height h,
I_0 = intensity of the incident radiation outside the earth's atmosphere,
σ = absorption cross-section,
h = height of observation, and
h_∞ = height of the top of the atmosphere.

At 44 to 60 A the absorption cross-sections for molecular oxygen and nitrogen are essentially the same. The atmosphere is treated as a single gas and the atmospheric density versus altitude is obtained from the intensities. If the cross-sections are different the gases are treated separately. Where the cross-sections are known and their atoms are strongly absorbing, the constituent densities may be found from the intensities of the lines as a function of altitude.

2.3.4 Mass spectrometers

Perhaps the most extensive measurements of the abundances of the different gases have been carried out with mass spectrometers. Early measurements were bothered by recombination of atomic oxygen to molecular oxygen at the spectrometer walls and surfaces. By jettisoning the nose cone Meadows and Townsend[12] with a radio-frequency spectrometer found gravitational separation of Ar and N_2.

Successful mass spectrometer O/O_2 ratios were measured by Pokhunkov[13] with a radio-frequency mass spectrometer on a rocket flight in September 1960. Schaefer and Nichols[14] in 1962 measured the O/O_2 ratio with a quadrupole mass spectrometer that had the advantages that the ion source was open to the atmosphere and that a negligible fraction of the gas in the analyzer could return to the source. Each molecule experienced only about one surface collision prior to ionization.

Excellent results were obtained by the Minnesota-Naval Research Group[15] with a single focusing permanent magnet spectrometer on a flight in June 1963. Their single and double focusing mass spectrometers are shown in Fig. 2.3. The O/O_2 and the N_2/O_2 ratios increased with height from 100 to 200 km as expected from the gravitational mass separation. The O/O_2 ratio was 15 at 210 km.

With double focusing magnetic spectrometers on the Explorer satellite Reber[16] found the He in the atmosphere that had earlier been predicted by Nicolet.[17] The analysis of drag data convinced Nicolet that He must be the major atmospheric constituent at a few hundred kilometers before H becomes predominant. In addition to He the

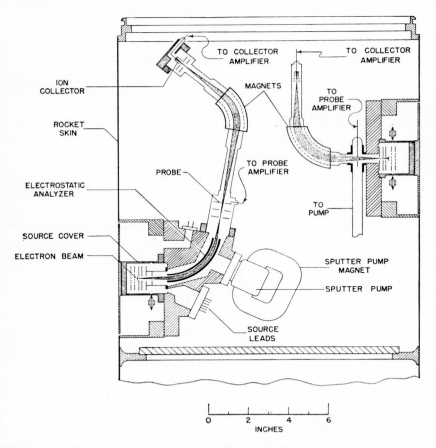

Figure 2.3. A cross section view of the single and double focusing mass spectrometers flown on a rocket in 1963. Taken from Nier, *et al.*,[15] 1964.

satellite experiment measured the abundances of O, N_2, and O_2 between 58 deg north and 58 deg south latitudes at altitudes up to 700 km. From 250 to 300 km N_2 and O were comparable. From 300 to 600 km atomic oxygen was the major constituent and above 600 km helium was predominant. The spectrometer was not set to measure the hydrogen abundance.

2.3.5 Atmospheric drag

Our knowledge of the atmospheric density from 200 to 3,500 km comes from atmospheric drag. It is similar to the dropping spheres method at lower altitudes discussed in section 2.3.2. It is the simplest, the most accurate and the most powerful method used to date. Every satellite is lowered by the force of air drag that changes the period of revolution of the satellite. The air density is then obtained from an accurate measurement of the period.

An excellent discussion of the methods of determining air density from satellite orbits is given by King-Hele.[18] Densities calculated from air drag avoid several difficult and costly operations of other methods, the design and construction of instruments to measure density, the design and building of facilities to test the instruments, and the launch of additional satellites. There are already many suitable satellites in orbit for observations. Only Newton's laws of motion and of gravitation and the molecular flow theory are needed to derive a value of the satellite drag coefficient. The resulting working equation is the drag Eq. 2.13 that was previously given for a sphere dropped from a rocket.

King-Hele[18] gives an example to demonstrate the simplicity of the drag method. Take the case of a nearly circular orbit with $\mathscr{F} = 1$ and the period of the orbit $\mathscr{P} = 2\pi a$. The a is the semimajor axis of the orbit and v is the satellite velocity. From Eq. 2.13

$$\rho = -\frac{0.106}{C_d\, a}\, \dot{\mathscr{P}}\, \frac{m}{A}.\qquad(2.15)$$

where $\dot{\mathscr{P}}$ is the time rate of change of the period.

Suppose a balloon satellite has a mass/area ratio m/A, of 0.3 g/cm². On the second and third days the satellite passes overhead at times delayed by 16 and 8 minutes from the time on the first day. The periods on the two days are 91.00 and 90.50 minutes. The time rate of change of the period is -0.5 min/day. Using Kepler's law, the mean period is 90.75 min and a is 6,690 km. Taking 6,370 km as the earth's radius, the average height of the satellite is 320 km. This gives the height at which the density applies. Take the drag coefficient as $C_d = 2.2$, give $\mathscr{P}$ in days/day and a in cm, then

$$\rho = 7.5 \times 10^{-15} \text{ g/cm}^3. \tag{2.16}$$

With an accuracy in time of 0.1 min the density is good to 2 per cent except for the drag coefficient uncertainty. With Baker–Nunn cameras 3,000 times the above accuracy can be obtained.

In an elliptical orbit the satellite meets more air drag near its perigee than at any other position: i.e., if the eccentricity is 0.1, the density is about 10^6 times higher at perigee than at apogee. The retardation near perigee causes the apogee point to walk in closer to the earth. But the perigee height remains almost constant. This is shown in Fig. 2.4a.

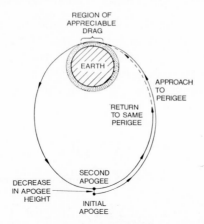

Figure 2.4a. A sketch of the contraction of a satellite orbit because of the drag of the earth's atmosphere at perigee. Taken from King-Hele,[18] 1966. The decrease in altitude at apogee is shown. The perigee height remains constant.

For a circular orbit the drag is spread out over the full 360 deg; however, if the eccentricity is 0.2 the drag is appreciable over a total arc of 40 deg. This significantly limits the spatial resolution of the drag measurements so that it is not possible to measure peaks in atmospheric density which are only a few degrees wide.

The contraction of the orbit is shown in Fig. 2.4b. Here the reductions in apogee are shown after 0.5 and 0.85 lifetimes. The perigee has

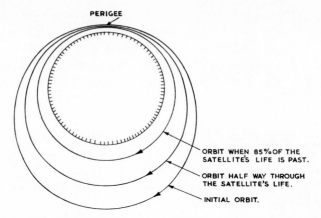

Figure 2.4b. A satellite's orbits shown at launch, halfway through the satellite's life, and 85 per cent through the satellite's life. Taken from King-Hele,[18] 1966. The perigee altitude has decreased only a small amount during this time.

changed very little during this time. Figure 2.5 shows the period of Explorer 1, 1958 α, the satellite that has monitored the atmospheric density near 350 km for more than 7 years. During this time the period has decreased from 115 min to 104 min.

The change in perigee height $h_{p0} - h_p$ in terms of orbit eccentricity is given[18] approximately by

$$h_{p0} - h_p \simeq \frac{H}{2} \ln \frac{e_0}{e}. \tag{2.17}$$

where e is the eccentricity and the subscript 0 refers to the initial conditions.

Take a perigee height of 200 km and $e = e_0/4$. It is observed experimentally that only 1/16 of the lifetime then remains. H is 25 km and from Eq. 2.17 $h_{p0} - h_p$ is found to be only 20 km. Using Kepler's law

$$\mathscr{P}/\mathscr{P}_0 = (a/a_0)^{3/2}, \tag{2.18}$$

and to the same approximation as in Eq. 2.18.

$$\frac{\mathscr{P}}{\mathscr{P}_0} = \left(\frac{1-e_0}{1-e}\right)^{3/2} \left(1 - \frac{3H}{4h_{p0}} \ln \frac{e_0}{e}\right) \tag{2.19}$$

One additional force that has not yet been taken into account is the rotation of the atmosphere. This force causes a small decrease in the

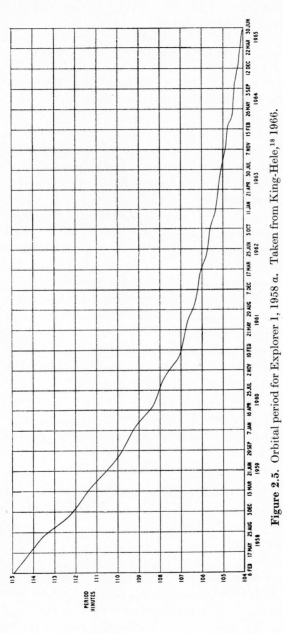

Figure 2.5. Orbital period for Explorer 1, 1958 *a*. Taken from King-Hele,[18] 1966.

angle of inclination i of the orbit to the equator. The decrease in i in degrees is given approximately by

$$i_0 - i \simeq 0.01 \; (\mathscr{P}_0 - \mathscr{P}) \sin i \qquad (2.20)$$

where $\mathscr{P}$ is given in min. As $\mathscr{P}$ decreases from 100 to 90 minutes near the end of the satellite's lifetime, i decreases by about 0.1 deg. If an accurate measurement of the decrease in i exists, on the other hand, it may be possible to obtain the rotation velocity of the atmosphere. This would be a useful method to obtain the atmospheric rotation wind.

The detailed formulae for determining air densities are given by King-Hele[18] and will not be reproduced here. However, some general comments about additional forces may be useful. The orbital period of the satellite is unaffected by the gravitation perturbations produced by a non-spherical earth and by the gravitational forces of the moon. This is because Kepler's third law, $\mathscr{P} \sim a^{3/2}$, is a function of a only. Since a is a measure of the satellite's total energy and since the gravitational forces are conservative, the orbit period is unaffected.

Radiation pressure can cause a change in the orbit period. The effect is usually negligible below 400 km, is appreciable and must be corrected for up to 1,000 km and becomes dominant above 1,000 km. However, if the satellite is continuously in the sun, i.e., a satellite with its orbit plane perpendicular to the earth–sun line, the effect can be neglected. Fea[19] used the balloon satellite 1963 30D to derive the hydrogen atmospheric density to 3,500 km. It appears that even at these heights the electric field drag forces can be ignored and it may be possible to extend the drag measurements to perhaps 5,000 km.

2.3.6 Other methods

Golomb, et al.[20] used the chemiluminescence of nitric oxide to measure the atomic oxygen from 90 to 140 km. A released jet of NO reacts with O

$$NO + O \rightarrow NO_2 + h\nu, \qquad (2.21)$$

and the light is observed photographically at ground stations.

Barth[21] flew a UV scanning spectrometer on a rocket flight from Wallops Island, Virginia in November 1963 to study the atmospheric airglow. The dayglow spectrum was obtained from 1500 to 3200 A. The main source of the UV light is the 1–0 γ band of NO at 2155 A.

A column density of 1.7×10^{14} NO molecules/cm² above 85 km was found.

2.4 ATMOSPHERIC DENSITY VARIATIONS

Before arriving at a best atmospheric density it is necessary to understand the time and space variations that occur. Excellent discussions of these variations have been written.[22] Variations with time and place are inevitable because of local atmospheric heating by the sun overhead. When the heat input increases at 100 to 200 km because of absorption of UV radiation from the sun the atmosphere rises and the atmospheric density increases at higher altitudes. Seasonal variations in heat input and variations over the solar cycle cause changes in the atmospheric density. Six variations can be identified: (1) day–night, (2) 11-year solar cycle, (3) 27-day sun rotation, (4) semi-annual, (5) geomagnetic activity, and (6) latitude.

The temperature contours caused by the local day–night heating are shown on the earth for the times of spring or fall and mid-summer in Fig. 2.6a and 2.6b.[23] Because of the sun's heat the observed densities reach a maximum at about 2:00 p.m. and a minimum at about 4:00 a.m. The atmospheric bulge has its maximum at 30 deg east of local noon and decreases uniformly on all sides. The hours of maxima and minima remain remarkably constant throughout the solar cycle.

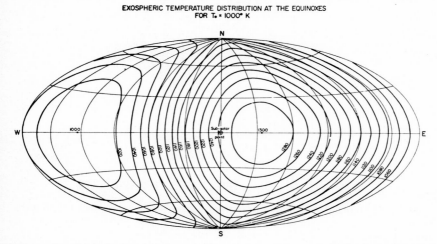

Figure 2.6a. Exospheric temperature contours in the spring and fall of 1964 for a midnight temperature of $1{,}000°$ K. Taken from Jacchia,[23] 1965.

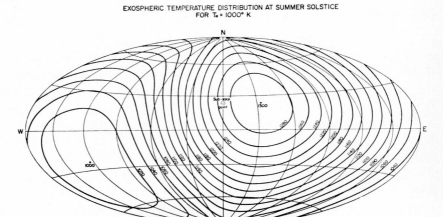

Figure 2.6b. Exospheric temperature contours at mid-summer for a midnight temperature of $1,000°$ K. Taken from Jacchia,[23] 1965.

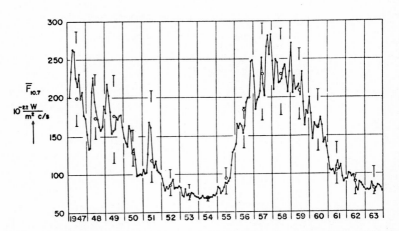

Figure 2.7. Monthly averages of the solar 10.7 cm flux in units of 10^{-22} W/m²–c/s from the National Research Council, Ottawa, Canada for the years from 1947 through 1963. The open circles represent the yearly averages. The bars indicate the scatter of the daily values, represented by the maximum and minimum values during the months of June and July for each year. Taken from CIRA,[30] 1965.

Nicolet[24] considers that the day–night variation is entirely due to UV heating. Harris and Priester[25] suggest a combination of UV and a second source, probably particles. King, *et al.*[26] think that ionic dis turbances alone are capable of explaining the daily variations without any direct help from UV.

Because the sun's energy output varies over the solar cycle, the absorbed energy in the atmosphere varies. The atmospheric tempera ture changes and this causes the atmospheric density to increase or decrease. The variation in density from solar minimum to maximum is more than a factor of 10 in the model of Harris and Priester at heights of a few hundred kilometers. The variation of the atmospheric density closely follows the 10.7 cm intensity as measured by the National Research Council, Ottawa, Canada. Data from 1947 to 1963 is shown in Fig. 2.7. The mean flux varied from about 65 to about 250×10^{-22} watts/m^2 – c/s over the last solar cycle.

Because the emission from some positions on the sun is greater than from others a 27-day atmospheric density variation due to the 27-day period of rotation of the sun is observed. The 10.7 cm radio intensities also follow the 27-day variations of solar activity. In a correlation study, Fig. 2.8, taken from Jacchia,[23] the atmospheric density and derived temperature at 350 km are compared to the 10.7 cm radio intensity and the geomagnetic index a_p. The semi-annual variation, the decrease in density with the 11-year solar cycle, and the 27-day variations are all clearly evident.

The close correlation of the atmospheric density with geomagnetic storms is shown in Fig. 2.9. There, the correlations of the atmospheric density with a_p are especially strong. The changes in the atmospheric density are more noticeable at high geomagnetic latitudes where the heating appears to be about two times greater. Increases of a factor of 3 to 10 in a matter of a few hours at times of large magnetic storms are seen. An example of a large increase in temperature correlated with a_p on October 28–29, 1961 is shown in Fig. 2.10. The increase in tempera ture with a_p can be expressed as

$$\Delta T \ (\mathrm{K}^\circ) \ = 1.0 a_p + 125 \ (1 - e^{-0.08 \ a_p}). \tag{2.22}$$

The mechanism of heating by magnetic storms is not understood. Possible explanations are Joule heating[27] and dissipation of hydro magnetic waves.[28] The density changes appear to be in phase at all height levels and to correspond to temperature changes just as for the

c

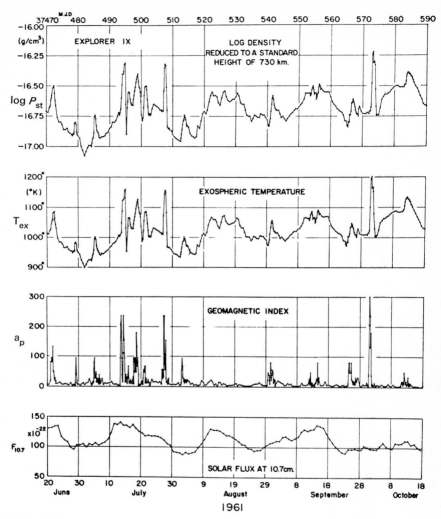

Figure 2.8. The atmospheric densities and temperatures derived from the drag on the Explorer 1 satellite, 1958 α. These are compared with the 10.7 cm solar flux and the geomagnetic index, a_p. Taken from Jacchia, [23] 1965.

UV radiation absorption. Therefore, the magnetic storm energy should be deposited at the same heights as the UV radiation.

The semi-annual atmospheric density maxima occur in April and October and the minima in January and July. The amplitudes of the variations are proportional to the 10.7 cm solar radio flux. The origin

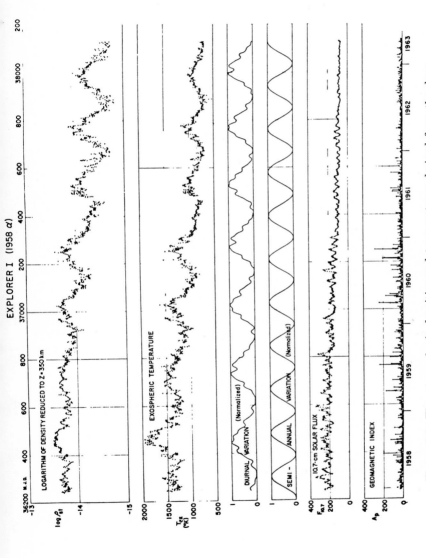

Figure 2.9. The atmospheric densities and temperatures derived from the drag of the Explorer 9 satellite, 1961 δ1. These are compared with the geomagnetic index, a_p, and the 10.7 cm solar flux. Taken from Jacchia, [23] 1965.

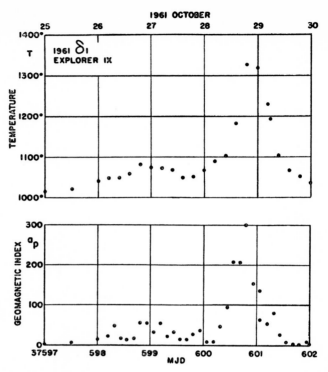

Figure 2.10. The atmospheric temperatures during a magnetic storm derived from the drag on Explorer 9, 1961 $\delta 1$, compared with the magnetic index, a_p. Taken from Jacchia,[23] 1965. MJD are Modified Julian Days—Julian Days minus 2,400,000.5.

of the semi-annual atmospheric variation is not known. It must be kept distinct from the semi-annual variation in the geomagnetic index. The seasonal variations could be accounted for if the daily bulge is elongated in the north–south direction and is permanently centered near the equator.[23] It is also possible that the temperature differences between the summer and winter poles cause convection currents in a north–south direction and produce the semi-annual variations.[29]

After removing the effects on drag due to daily, 27-day, 11-year, geomagnetic and semi-annual variations, an effect still remains which might be a latitude effect.

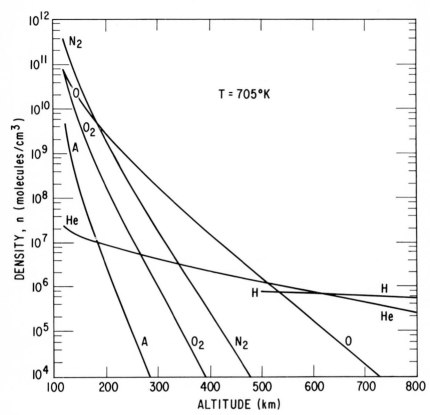

Figures 2.11a, 2.11b, and 2.11c. The atmospheric densities for molecules at altitudes of 100 to 800 km for temperatures of 705° K, 1,320° K, and 1,970° K. Taken from CIRA,[30] 1965. These are typical of the minimum, average, and maximum heating during the solar cycle.

2.5 ATMOSPHERIC COMPOSITION

The measurements outlined in Section 2.4 have been combined with the theory of gravitational separation by different authors to propose model atmospheres. The latest extensive effort is the CIRA 1965 atmosphere.[30] There, densities incorporating the daily and solar cycle variations up to 800 km are tabulated. For the atmospheric composition and density here we incorporate the following models: (1) up to 10 km

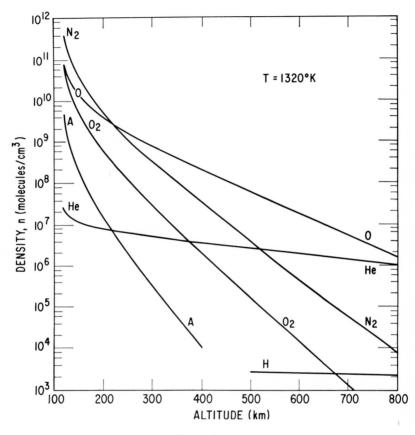

Figure 2.11b

the composition given by List,[31] (2) from 10 to 120 km the model of Sissenwine,[32] (3) from 120 to 800 km the CIRA 1965 atmosphere of Harris and Priester,[33] (4) from 800 to 3,500 km the drag data of Fea,[19] and (5) above 3,500 km the theoretical calculations of Johnson.[34]

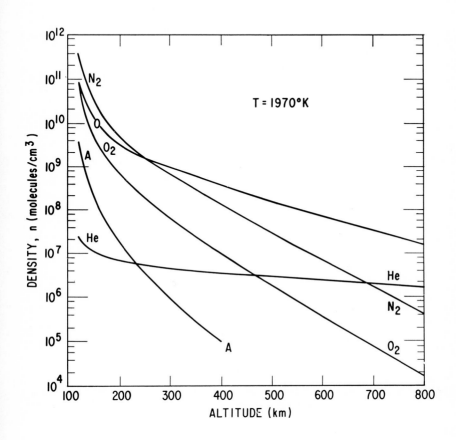

Figure 2.11c

The composition of the atmosphere in Figs. 2.11a, 2.11b and 2.11c is taken from the CIRA 1965 atmosphere and is given for three temperatures, 1970° K, 1320° K and 705° K, that correspond to different amounts of heating of the atmosphere. These should be typical of maximum, intermediate, and minimum heating during the solar cycle.

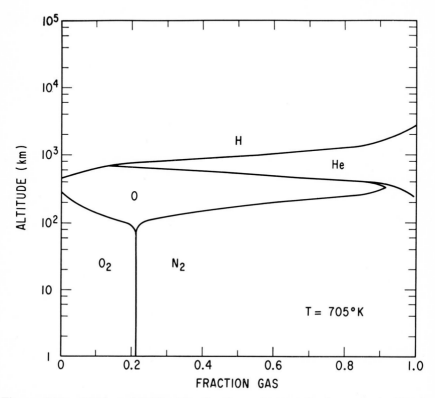

Figures 2.12a, 2.12b, and 2.12c. The fractions of atmospheric gases at altitudes from 1 to 10^5 km for temperatures of 705° K, 1,320° K, and 1,970° K. The fraction of a given gas is found from the horizontal distance on the graph between its two boundaries.

It is perhaps easier to visualize the composition of the atmosphere as the fraction of the total atmosphere that each gas contributes. These "phase-like" diagrams are given in Figs. 2.12a, 2.12b and 2.12c. The fraction of gas, at a particular altitude, is found by the distance along the horizontal scale between the two boundary lines of the gas. From these plots it can be seen that N_2 is always more abundant than O_2, and is the

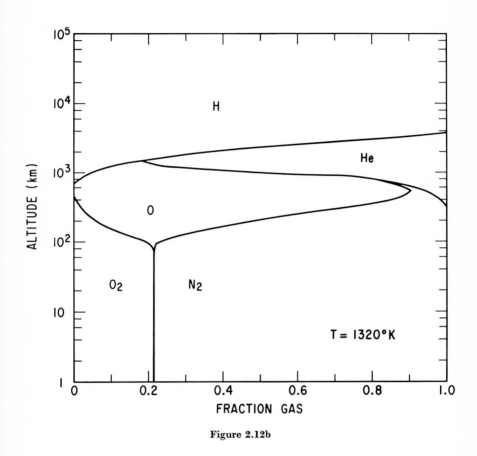

Figure 2.12b

most abundant gas up to 225 km. Molecular oxygen is broken up into atomic oxygen by the sun's UV radiation at altitudes above 100 km. Atomic oxygen is therefore the most abundant gas from 200 to 500 km when the heat absorption is the lowest at the minimum of the solar cycle and at night. When the heat absorption is the greatest, at solar maximum during the day, oxygen is the most abundant gas from 250 to

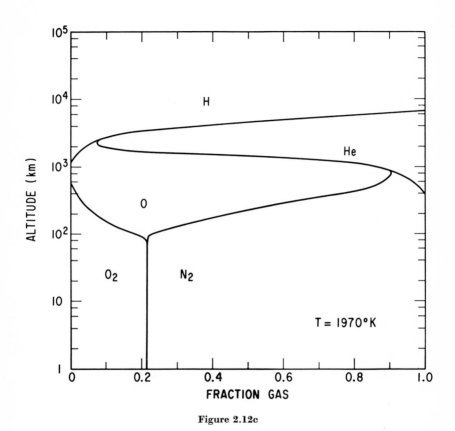

Figure 2.12c

1,500 km. At the time of the least heating, He is most abundant from about 500 to about 1,000 km and during the greatest heating is most abundant from about 1,500 to 5,000 km. This leaves H as the major constituent above 1,000 km for minimum heating and above 5,000 km for maximum heating.

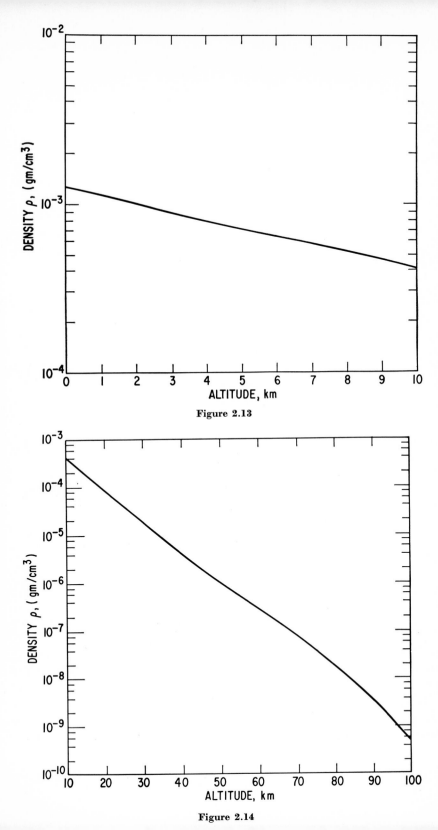

Figure 2.13

Figure 2.14

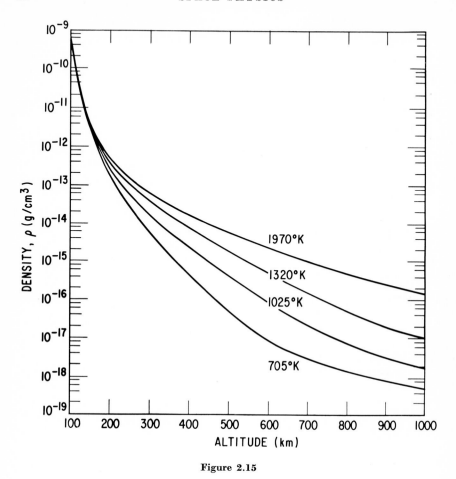

Figure 2.15

Figures 2.13, 2.14, 2.15, 2.16, and 2.17. The atmospheric densities versus altitude from the surface of the earth to 100,000 km. These densities incorporate the following models: (1) up to 10 km the composition given by List,[31] 1957; (2) from 10 to 120 km the model of Sissenwine,[32] 1962; (3) from 120 to 800 km the CIRA 1965 atmosphere of Harris and Priester,[33] 1965; (4) from 800 to 3,500 km the drag data of Fea,[19] 1965, and (5) above 3,500 km the theoretical calculations of Johnson,[34] 1962. The densities are not temperature dependent below 100 km. Above 100 km, densities for temperatures from the minimum to the maximum heating are given. Note the crossover of the density dependence on temperature at a few thousand kilometers. The reader is cautioned that little experimental data exists above 3,500 km and that the values on the graphs are theoretical and contain considerable uncertainty.

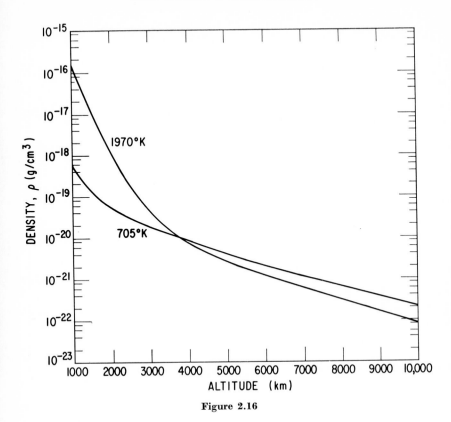

Figure 2.16

2.6 ATMOSPHERIC DENSITY

The atmospheric constituents of Section 2.4, derived from the models for different altitude regions, are combined to obtain the atmospheric densities. It is necessary to merge smoothly across the transition regions from one model to the next. The densities are presented in five altitude regions: 0–10, 10–100, 100–1,000, 1,000–10,000 and 10,000–100,000 km in Figs. 2.13, 2.14, 2.15, 2.16, and 2.17.

The reader is cautioned that there is little experimental data above 3,500 km and that the values on the curves are theoretical, only. The values depend on the rate of escape of H from the atmosphere and could change significantly with the addition of new measurements. Even at the low altitudes the densities can be expected to change by as much as a factor of 2 as new and better measurements become available. Recent analyses[35] of drag from Cosmos satellites at altitudes down to 170 km

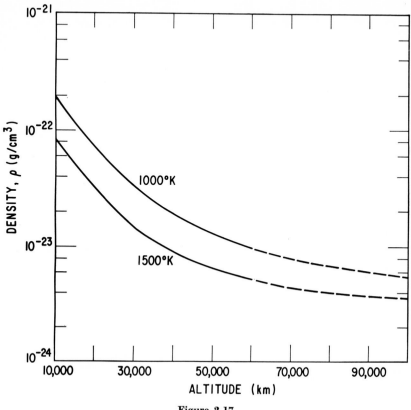

Figure 2.17

indicate significant departures from the CIRA 1965 model atmosphere. In 1965 the observed day–night variation was more than a factor of 2 at 200 km and was still observable at 170 km.

The measurements by different methods are not always in agreement. The drag measurements on Explorer 17 were generally higher than the ionization gage measurements by a factor of 2.[36] This disagreement has not been resolved.

For purposes of computing the energy loss of charged particles moving in the earth's magnetic field it is necessary to average the atmospheric densities over the particle's motion around the earth. The average is taken over the bounce and drift motions of the particles. Usually the circular cyclotron orbit can be neglected. Only for protons with energies greater than about 100 MeV at altitudes less than about

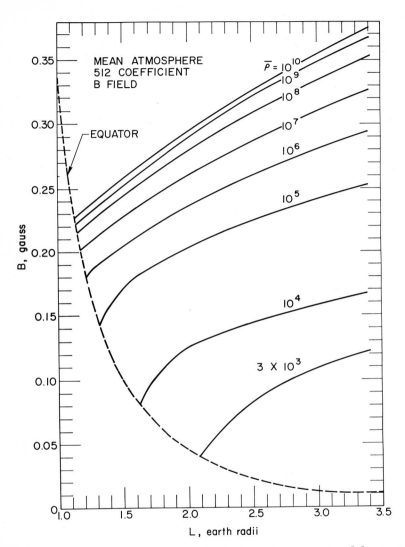

Figure 2.18. Contours of mean atmospheric densities encountered by trapped protons on a magnetic field versus magnetic shell parameter, B–L, graph. Taken from Cornwall, *et al.*[37] 1965. The densities are given in atomic electrons/cm³. The atmospheric models presented here and a 512 coefficient magnetic field representation were used to compute the mean densities.

300 km is the difference in the atmospheric density over the cyclotron orbit enough to require a correction.

In order to perform the average it is necessary to use a model for the atmospheric density and a representation for the earth's magnetic field. Since the atmospheric density changes over the solar cycle, the particular time in the solar cycle for the calculation must be specified. Or if a mean over the solar cycle is desired a further average must be made. The mean density [37] averaged over the solar cycle with a 512 coefficient representation of the earth's magnetic field and a modified Harris–Priester model of the atmosphere is reproduced in Fig. 2.18. Contours of equal atmospheric density are presented on a B–L plot where B is the value of the magnetic field and L is the magnetic shell number. See Chapt. 1, section 1.1. This method of presentation is useful for experimenters who analyze their data using B–L plots.

It is emphasized that there is a rather large uncertainty in this mean atmospheric density calculation. At low altitudes the calculation is sensitive to the magnetic field representation and at high altitudes the atmospheric densities have not been measured. In view of these uncertainties it is possible that mean densities could have errors as large as a factor of 10.

2.7 CONCLUSIONS

Great progress has been made in the understanding of the atmosphere since the launching of the first satellite in 1957. Information has been added by experiments on both rockets and satellites. Gravitational separation of atmospheric gases above 120 km has been established. The composition of the atmosphere has been measured up to 800 km and drag measurements of densities have been extended up to 3,500 km. Day–night variations of a factor of 10 and solar cycle variations greater than a factor of 10 have been observed. In addition, there are smaller 27-day variations corresponding to the sun's period of rotation and also semi-annual variations that are not understood. Variations correlated with magnetic storms are regularly seen. There is some evidence for latitude variations apart from other affects.

Many questions still must be answered. Is the solar UV the only source of atmospheric heating? Or is significant heating also caused by the penetrating high energy charged particles or by the slowly drifting low energy charged particles? Why are variations of the earth's magnetic field accompanied by heat effects? What causes the semi-annual variations in the atmospheric density? Measurements over small distances in latitude or longitude that could search for atmospheric density spikes have not been made. Measurements have not been taken that could detect density changes in short times. And finally, what really is the atmospheric composition and density above 1,000 km where few measurements exist?

REFERENCES

1. Willet, H. C. and Sanders, F., *Descriptive Meteorology*, Academic Press Inc., New York, p. 51–54 (1959).
2. Nordberg, W., "Geophysical observations from NIMBUS I," *Science* **150**, 559–572 (1965).
3. Johnson, F. S., "Structure of the upper atmosphere," *Satellite Environment Handbook, Revised*, Stanford University Press, 1964.
4. Champion, K. S. W., *CIRA 1965 COSPAR international reference atmosphere*, p. 7, North-Holland Publishing Co., Amsterdam, 1965.
5. Ching, B. K. and Becker, R. A., "Density, temperature, and neutral composition of the atmosphere above 100 km," *Aerospace Corporation Report* No. TDR-469(5260-10)-1, 1965.
6. Havens, R. J., Koll, R. T. and LaGow, H. E., "The pressure, density, and temperature of the earth's atmosphere to 160 km," *J. Geophys. Res.* **57**, 59–72 (1952).
7. LaGow, H. E., Horowitz, R. and Ainsworth, J., "Results of I.G.Y. atmospheric density measurements above Fort Churchill," *Space Research I*, p. 164, North-Holland Publishing Co., Amsterdam, 1962.
8. Faucher, G. A., Procunier, R. W. and Sherman, F. S., "Upper-atmosphere density obtained from measurements of a drag on a falling sphere," *J. Geophys. Res.* **68**, 3437–3451 (1963).
9. Horowitz, R., "Direct measurements of density in the thermosphere," *Ann. Geophys.* **22** (1), 31–39 (1966).
10. Byram, E. T., Chubb, T.A. and Friedman, H., "Dissociation of oxygen in the upper atmosphere," *Phys. Rev.* **98**, 1594–1597 (1955).
11. Hinteregger, H. E., "Absorption spectrometric analysis of the upper atmosphere in the EUV region," *J. Atmos. Sci.* **19**, 351–368 (1962).
12. Meadows, E. B. and Townsend, Jr., J. W., "Diffusive separation in the winter nighttime arctic upper atmosphere 112 to 150 km," *Ann. Geophys.* **14**, 80–93 (1958).
13. Pokhunkov, A. A., "Gravitational separation, composition, and structural parameters of the night atmosphere at altitudes between 100 and 210 km," *Planet. Space Sci.* **11**, 441–449 (1963), translated from *Iskusstvennue Sputniki, Zamli*, No. **13**, p. 110, 1962; "Mass-spectrometric investigation of the neutral composition of the upper atmosphere," *Ann. Geophys.* **22** (1) 92–101 (1966).
14. Schaefer, E. J. and Nichols, M. H., "Neutral composition obtained from a rocket-borne mass spectrometer," *Space Research IV*, p. 205, North-Holland Publishing Co., Amsterdam, 1964.

15. Nier, A. O., Hoffman, J. H., Johnson, C. Y. and Holmes, J. C., "Neutral composition of the atmosphere in the 100–200 km range," *J. Geophys. Res.* **69**, 979–989 (1964); Nier, A. O., "The H neutral composition of the thermosphere," *Ann. Geophys.* **22** (1), 102–109 (1966).

16. Reber, C., "Data from Explorer 17 on composition of the upper atmosphere," *J. Geophys. Res.* **69**, 4681–4685 (1964).

17. Nicolet, M., "Helium, an important constituent in the lower exosphere," *J. Geophys. Res.* **66**, 2263–2264 (1961).

18. King-Hele, D. G., "Methods of determining air density from satellite orbits," *Ann. Geophys.* **22** (1), 40–52, (1966).

19. Fea, K., "Determination of the density of air at an altitude of 3,500 km," *Nature* **205**, 379–382, Jan. 23, 1965; "Exospheric conditions, to a height of 3,500 km, derived from satellite accelerations in 1964," *Planetary and Space Science* **14**, 291–297 (1966).

20. Golomb, D., Rosenberg, N. W., Aharonian, C., Hill, J. A. F. and Alden, H. L., "Oxygen atom determination in the upper atmosphere by chemiluminescence of nitric oxide," *J. Geophys. Res.* **70**, 1155–1174 (1965).

21. Barth, C. A., "Rocket measurement of nitric oxide dayglow," *J. Geophys. Res.* **69**, 3301–3303 (1964).

22. Jacchia, L., "Density variations in the heterosphere," *Ann. Geophys.* **22** (1), 75–85 (1966); Priester, W., "On the variations of the thermospheric structure," *Proc. Roy. Soc.* **A288**, 493–509 (1965).

23. Jacchia, L., "Atmospheric structure and its variations at heights above 200 km," *CIRA* **1965** *COSPAR International Reference Atmosphere*, p. 293, North-Holland Publishing Co., Amsterdam (1965).

24. Nicolet, M., "Solar radio flux and temperature of the upper atmosphere," *J. Geophys. Res.* **68**, 6121–6144 (1963).

25. Harris, I. and Priester, W., "Time dependent structure of the upper atmosphere" *Inst. for Space Studies, Goddard Space Flight Center, Natl. Aeronautics and Space Admin. Report* X-640–62–69; "Theoretical models for the solar-cycle variation of the upper atmosphere," *Inst. for Space Studies, Goddard Space Flight Center, Natl. Aeronautics and Space Admin. Report* X-640–62–70 (1962).

26. King, J. W., Eccles, D., Legg, A. J., Smith, P. A., Galindo, P. A., Kaiser, B.A., Preece, D. M. and Rice, K. C., "An explanation of various ionospheric and atmospheric phenomena including the anomalous behavior of the F-region," *Radio Research Station, Ditton Park, Slough, England, Document* No. RRS/I.M. 191, December (1964).

27. Cole, K. D., "Joule heating of the upper atmosphere," *Australian J. Phys.* **15**, 223–235 (1962).

28. Dessler, A. J., "Ionospheric heating by hydromagnetic waves," *J. Geophys. Res.* **64**, 397–401 (1959).

29. Johnson, F. S., "Circulation at ionospheric levels," unpublished (1964).

30. "*CIRA 1965 COSPAR International Reference Atmosphere*," North-Holland Publishing Co., Amsterdam, 1965.

31. List, R. J., *Smithsonian Meteorological Tables, Sixth Revised Edition* p. 291–294 and 389 (1951).

32. Sissinwine, N., "Announcing the U.S. standard atmosphere—1962," *Astronautics* **52-53**, August (1962).

33. Harris, I. and Priester, W., "*CIRA 1965 Cospar International Reference Atmosphere*, p. 91–260, North-Holland Publishing Co., Amsterdam, 1965;

"Relation between theoretical and observational models of the upper atmosphere," *J. Geophys. Res.* **68**, 5891–5894 (1963).

34. Johnson, F. S., "Atmospheric structure," *Astronautics*, 54–61 August (1962).
35. King-Hele, D. G. and Quinn, E., "Upper atmosphere density, determined from orbits of Cosmos rockets," *Planet. Space Sci.* **14**, 107–111 (1966).
36. Newton, G. P., Horowitz, R. and Priester, W., "Atmospheric densities from Explorer 17 density gauges and a comparison with satellite drag data," *J. Geophys. Res.* **69**, 4690–4692 (1964).
37. Cornwall, J. M., Sims, A. R. and White, R. S., "Atmospheric density experienced by radiation belt protons," *J. Geophys. Res.* **70**, 3099–3111 (1965).

CHAPTER 3

The Earth's Ionosphere

3.1 INTRODUCTION

The earth is surrounded by electrons and charged atoms and molecules called ions that make up the ionosphere. They are in addition to the neutral molecules and atoms of the atmosphere. The number of positively charged ions in the ionosphere is equal to the number of electrons plus negatively charged ions. Otherwise the excess of one kind of charge would set up electric fields that would cause charges to flow in the conducting ionosphere and would bring regions of excessive positive or negative charges quickly to charge neutrality. There is some speculation that small fields can be maintained in certain regions of the magnetic cavity. This will be discussed in section 3.5 and in Chapter 6.

The electrons and ions of the ionosphere collide and interact with the molecules and atoms of the neutral atmosphere. In this way, the temperatures of the electrons and ions are held close to the temperatures of the neutral particles, particularly at the low altitudes. At the higher altitudes where collisions are less frequent the thermal contact is not so good and the electron temperatures are often higher than the temperatures of the ions, molecules, and atoms.

The electrons and ions of the ionosphere have much lower energies than the radiation belt trapped electrons and protons that were discussed in Chapter 1. The trapped radiation belt particles have energies that vary from a few eV to a few hundred MeV but the energies of the ionospheric electrons and ions are usually only a fraction of one eV. The number densities of the particles in the ionosphere vary from one to 10^6 particles/cm³ while the trapped protons and electrons seldom exceed one particle/cm³.

Both ionosphere and trapped radiation belt particles are limited by the magnetic field energy density

$$E = \frac{B^2}{8\pi} \text{ ergs/cm}^3. \tag{3.1}$$

For a magnetic field B, of 0.3 gauss at the equator near the earth $E = 3.6 \times 10^{-3}$ erg/cm^3 or 2×10^9 eV/cm^3. This energy density is the equivalent of 2×10^9 ionospheric particles/cm^3 of 1 eV each or 2×10^4 trapped radiation belt particles/cm^3 of 100 keV each. Since the magnetic field falls off as $1/R_e^3$, at 5 earth radii the limit to the number density is reduced to 2×10^7 ionospheric particles/cm^3 or 200 trapped particles/cm^3. This magnetic field limit is never approached close to the earth but is effective in limiting the particle densities at large distances from the earth where the magnetic field is weak.

The discovery of the ionosphere followed Marconi's wireless transmission from England to the United States in 1909. The radio wave energy travelled much farther than permitted by the inverse square law. The energy should have been lost out into space. Earlier in 1902 Kennelly[1] and Heaviside[2] had suggested independently that an ionized layer around the earth could guide the energy great distances. Kennelly pointed out that the energy should fall off linearly with distance instead of as the inverse square. This ionized layer affecting the radio waves has since been named the Kennelly–Heaviside layer.

The height of the Kennelly–Heaviside layer and its ionization and its properties were measured in a beautiful set of experiments by E. V. Appleton and his associates. For this pioneering work Appleton received the Nobel prize in physics in 1947. His early experiments and theoretical work are reviewed in a lecture[3] delivered before the Wireless Society in May, 1932. In this lecture he discussed the ionosonde method of determining the ionospheric density. It furnished much of our knowledge of the ionosphere until the experiments on rockets in the late 1940's and the satellites in the late 1950's.

Some of the leading physicists of the early twentieth century contributed to the explanation of radio wave refraction. H. A. Lorentz[4] introduced his theory of the propagation of light waves through a layer of molecules in 1909, W. H. Eccles[5] proposed the theory of wave propagation by ions in the ionosphere in 1912 and J. Larmor[6] emphasized the importance of free electrons in the refraction of radio waves. When the radio waves travel through the electrons of the ionosphere their phase velocity increases. The index of refraction is less than one and is given by

$$\mu^2 = 1 - \frac{4\pi N_e\, e^2}{m_e \omega^2}.$$ (3.2)

N_e is the electron density in electrons/cm³, e is the electron charge 4.8×10^{-10} e.s.u., m_e is the mass of the electron 9.1×10^{-28} gm, and $\omega = 2\pi f$ where f is the radio wave frequency. A second term $a4\pi N_e\, e^2$ could be added to the $m_e \omega^2$ in the denominator to take into account local polarization effects and interelectronic influences. However, we will neglect these effects and take $a = 0$.

The transmission of the wave into media with different indices of of refraction can be treated just as for other electromagnetic waves, light for example. When the wave enters a different medium it is refracted, it changes its direction and the wave front tips. In the iono-sphere the wave gradually bends around until it reverses its direction and returns to earth. This is seen in Fig. 3.1. The wave progresses from the transmitter on the earth at a, through the directions $bcde$ and returns to the receiver on the earth at f. The angle of incidence under the ionosphere is i and the angle of refraction at any position c is r.

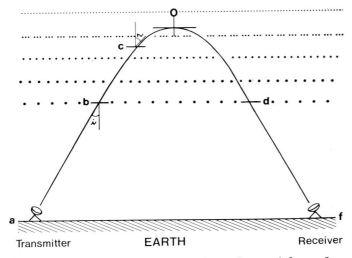

Figure 3.1. Radio wave refraction from the ionosphere. A beam from a radio transmitter at a on the ground enters the ionosphere at b and is refracted by the increased electron density at the higher altitudes. The angle of incidence i is shown at b and the angle of refraction r at c. The angle of refraction at O is 90 deg. The beam returns from its maximum height at O to the receiver on the earth at f

The index of refraction outside the ionosphere is μ_b, and at c is μ_c. We know from the optics law of refraction that

$$\mu_b \sin i = \mu_c \sin r. \tag{3.3}$$

In the rare atmosphere under the ionosphere μ_b is one. At the top of the trajectory the angle $r = 90$ deg so $\sin r = 1$. Then $\mu_0 = \sin i$. For the case where the wave rises nearly vertically under the ionosphere, angle $i = 0$ and $\mu_0 = 0$. Setting $\mu^2 = 0$ in Eq. 3.2 and $\omega_0 = 2\pi f_0$, we find

$$f_0 = \sqrt{\frac{N_e\, e^2}{\pi m_e}}. \tag{3.4}$$

This is the cutoff frequency. All frequencies less than f_0 are returned to earth. The higher frequencies penetrate the ionosphere and go out into space. Substituting for the constants, the ionospheric density at the tangent point is

$$N_e = 1.24 \; 10^{-8} f_0^2 \; \text{electrons/cm}^3 \tag{3.5}$$

Actually the refractive index is modified by the earth's magnetic field B. Two refractive indices are possible depending upon the direction of the wave normal with respect to the magnetic field. The two waves are called the ordinary and the extraordinary waves. The ordinary wave is not affected by the magnetic field and its frequency is given by Eq. 3.4. However, the extraordinary wave frequency is modified to be

$$f_x = \frac{f_{Be}}{2} \sqrt{\frac{f_{Be}^2}{4} + f_0^2}, \tag{3.6}$$

where the electron cyclotron frequency f_{Be}, is given by

$$f_{Be} = \frac{Be}{2\pi\, m\, c}. \tag{3.7}$$

Appleton used a frequency-change method to determine the height of the ionospheric refraction. The time differences for the wave to travel from the transmitter to the receiver via the ionosphere and via the ground were measured at different frequencies. Then, using the distance between the transmitter and receiver he calculated the height.

A widely used method of measuring the atmospheric density was developed by Breit and Tuve.[7] A short time burst of radio waves is transmitted at one frequency. The waves penetrate through the ionosphere up to the height where the ionospheric electron density satisfies

Eq. 3.5. Then the waves are returned. The electron density is calculated from Eq. 3.5. The height of the reflection is obtained from the time t that the burst takes in its travel to the ionosphere and back. For a vertical wave, the height h of the ionization is obtained from

$$h = \frac{ct}{2}, \tag{3.8}$$

where c is the velocity of light. A range of radio frequencies are swept to cover the entire range of electron densities. It is not possible to obtain electron densities at altitudes higher than the position of the maximum electron density by this method. The correct frequencies to satisfy Eq. 3.5 were already bent back to earth by the electron densities at lower altitudes.

From Eq. 3.5 a density of 10^4 electrons/cm^3 gives a cutoff frequency of 10^6 cycles/sec. A height of 100 km requires a travel time of about 10^{-3} sec so the pulse length of the burst must be much shorter than this.

A historical review by Kirby, Berkner, and Stuart[8] discusses the observations of the Bureau of Standards and summarizes ionospheric radio research to 1934. Three major daytime ionization layers had been located, an E layer at 100 to 200 km, an F_1 layer at about 180 km and an F_2 layer at about 240 km. The variations during the day and over the seasons were measured. The peaks in the atmospheric densities of the E and F_1 layers were found to occur at local noon independent of the seasons with the maximum intensities occurring in the summer. The densities in the F_2 layer were greatest near noon in the winter but after sunset in the summer. And the values of the F_2 layer densities varied widely from hour to hour and from day to day while the values of the E and F_1 layers varied only slightly. The electron densities occasionally reached a maximum of 2.5×10^6 electrons/cm^3 in the evening.

During World War II many ionospheric stations were established around the world. These were useful in discovering ionospheric latitude and longitude effects. One of the most unexpected results was the equatorial anomaly.[9] During the spring and fall when the sun shines on the equator the cutoff frequencies for the F_2 layer f_{F2}, were found to have two peaks displaced from the equator. One was at 30 deg north and the other at 30 deg south of the equator. The electron density was not maximum at the equator as might have been expected at first thought. The anomaly is shown in Fig. 3.2 where f_{F2} (is plotted versus

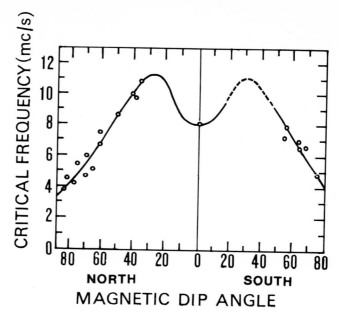

Figure 3.2. The variation of the cut off frequency f_{F2} with magnetic dip angle taken from Appleton.[9] The anomalous dip at the equator, observed in March 1944,[9] is symmetric about the magnetic equator.

the magnetic dip angle, the angle between the local magnetic field and the horizontal to the earth's surface. Considerable effort has subsequently been devoted to the measurement and understanding of this effect.

Lightning as a natural source of radio waves was discovered by Barkhausen[10] in 1919. Lightning flashes are heard as clicks on a receiver tuned between 15 kc/sec and 20 mc/sec. At the lower sound frequencies the tone consists of a steadily falling pitch. It sweeps several octaves in a few seconds. This radio noise is called a "whistler." Whistlers were studied early in 1933 by Eckersly[11] in England and by Burton and Boardman[12] in the United States. They showed that at time Δt after a loud click, the frequency f of the whistler could be written as

$$f^{-1/2} = \Delta t/D \qquad (3.9)$$

D is called the dispersion of the whistler.

A review of whistlers up to 1953 is given by L. R. O. Story.[13] He explained the differences between the whistlers with and without atmospheric clicks, called "long" and "short" whistlers. The click is produced by lightning within 2,000 miles that travels directly to the receiver. Clicks from lightning from distant points like the opposite hemisphere cannot be heard. The multiple long whistlers arrive at times related by 1:2:3:4:··· because they travel to the opposite hemisphere and back and each additional trip adds one unit of time. The short whistlers with no clicks are separated by times of 1/2:1 1/2:2 1/2:3 1/2···· They originate in the opposite hemisphere so the first trip is only one-half as long while the subsequent trips are from the receiver hemisphere to the opposite hemisphere and back to the receiver. A diagram of the paths followed by the short and long whistlers is sketched in Fig. 3.3. The greater production of long

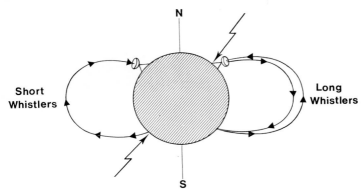

Figure 3.3. Paths of short and long whistlers in the earth's magnetic field. A short whistler originates from lightning in the opposite hemisphere from where it is detected. The long whistler originates in the same hemisphere, goes to the opposite hemisphere and returns. Taken from Storey,[13] 1953.

whistlers in the summer than in the winter is caused by the seasonal variation of thunderstorms. The accompanying decrease in the number of short whistlers is explained by the reversal of seasons in the opposite hemisphere.

The whistler wave packets are guided through the ionosphere by the magnetic field lines. Even if the waves are incident on the base of the ionosphere at large angles, the wave normals are soon refracted into a narrow cone about the vertical direction. The ray directions, the directions of energy propagation, closely follow the earth's magnetic field directions. And the lateral spread of the rays about the path of the central ray is small.

We follow Bourdeau[14] and divide the ionosphere into four regions shown in Fig. 3.4. These are the D region from 50 to 85 km altitude, the E region from 85 to 140 km, the F region from 140 to 600 km and the high ionosphere from 600 km to the edge of the earth's magnetic cavity. We will start with the D region, the one closest to the earth.

3.2 D REGION

The D region corresponds closely to the atmospheric mesosphere. Here the collision rate between the electrons and the neutral atoms and

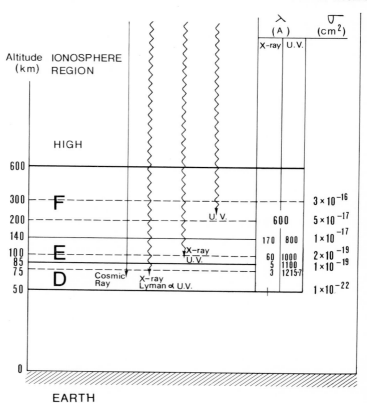

Figure 3.4. The regions of the ionosphere. The ionosphere is divided into the D, E, F, and High regions. The altitudes of the boundaries of the regions are shown at the left. The main radiations responsible for ionization in the different regions are indicated. Cosmic rays ionize in the low D region, X-rays and Lyman-α UV in the E region and UV in the F region. The wavelengths of the absorbed electromagnetic radiation and their cross sections are shown on the right-hand side of the figure.

molecules is high. The electrons quickly transfer their energy to the neutrals and come into temperature equilibrium.

Cosmic rays are the source of the electrons below 70 km in the D region and X-rays and Lyman-α ultraviolet radiation are the sources at the higher altitudes.[15] (See Fig. 3.4.) Visible light is transmitted through the region without absorption. Under solar quiet conditions, only cosmic rays have sufficient energies and low enough cross sections to penetrate below 70 km. They are mostly protons at relativistic velocities. They leave 5×10^{-3} ion-electron pairs/cm of path length at 70 km. A cosmic ray flux of 2 protons/cm²-sec at the equator gives an electron source of 10^{-2} electrons/cm³-sec.

At the altitudes of 70 to 85 km in the upper D region, two sources appear equally important, X-rays of 2 to 8 A (5 to 1.5 keV) with intensities of 10^{-3} erg/cm²-sec[16] and Lyman-α UV of 1215.7 A (10 eV) with an intensity of 3 to 6 erg/cm²-sec.

X-rays produce electrons by the photoelectric effect and by Compton scattering. The photoelectric effect is most important at low energies. X-rays are absorbed by atmospheric atoms and molecules and an electron is emitted with the energy of the X-ray less the ionization potential. At energies greater than 10 keV (1 A) the Compton effect becomes important. The X-ray gives up part of its energy to an electron and scatters off with reduced energy.

The X-ray reactions are

$$\text{X-ray} + O_2 \rightarrow O_2^+ + e \tag{3.10}$$

and

$$\text{X-ray} + N_2 \rightarrow N_2^+ + e \tag{3.11}$$

to form O_2^+ and N_2^+ ions and electrons.

Lyman-α UV reacts with nitric oxide to form NO^+ and an electron

$$\text{Ly-}\alpha + NO \rightarrow NO^+ + e. \tag{3.12}$$

The cross section for this reaction is very high; therefore, the concentration of NO in the atmosphere can be very low. In order to verify the significance of this reaction it is important to measure the concentration of NO throughout the atmosphere. To date it has not been done.

The number of electrons/cm³-sec N_e, produced by n_λ photons/cm²-sec of wavelength λ incident upon N_j neutral atoms or molecules of a particular kind j is

$$N_e = n_\lambda N_j \sigma_{\lambda j}. \tag{3.13}$$

$\sigma_{\lambda j}$ is the cross section in cm² for the particular reaction. It is necessary to sum over the sun's photon distribution at the top of the atmosphere, and over the different atoms and molecules that make up the atmosphere. The attenuation of the photons from the top of the atmosphere to the altitude of interest must also be included.

The time variation of the electron density may be estimated from

$$dN_e/dt = q - \alpha N_e^2, \tag{3.14}$$

where q is the number of electrons/cm³-sec produced and α cm³/sec is the recombination coefficient. It has been assumed that the number of electrons equals the number of positive ions. In the D region, since the recombination time is short, ionization equilibrium can be determined by setting dN_e/dt equal to 0. Then $N_e^2 = q/\alpha$. It is necessary to sum over the source and recombination coefficients for the atmospheric constituents involved.

The dominating loss mechanism for electrons in the D region appears to be dissociative recombination of the electrons on positive ions M⁺. The neutral molecule or atom is then left in an excited state,

$$e + M^+ \rightarrow M^*. \tag{3.15}$$

At times of solar flares the ionization often increases by orders of magnitude and the maximum in the D layer is reduced by 10 km in height. Since the Lyman-α radiation is nearly constant, even in large flares, and since X-rays of wavelengths less than 8 A change by orders of magnitude, the increased ionization in the D layer must be due to the X-rays. The SR-1 satellite observations[16] normally gave a flux of less than 0.6×10^{-3} erg/cm²-sec at wavelengths below 8 A. However, at times of solar disturbances this X-ray flux increased to more than 1.3×10^{-2} erg/cm²-sec.

In a polar cap absorption event the intensity of cosmic VHF radio waves is reduced. This is due to additional ionization that is produced by the solar protons that penetrate to the D region.[17] In addition, there is an absorption of HF radiation at auroral latitudes. High auroral optical activity is accompanied by a high amount of ionization. Electrons that are incident on the atmosphere have sufficiently high energies to penetrate to the D region and cause this ionization.

In Fig. 3.5 the electron density in the D region is given versus altitude for average conditions, a sudden ionospheric disturbance, an auroral absorption, and a polar cap absorption.[14] The ionization sometimes

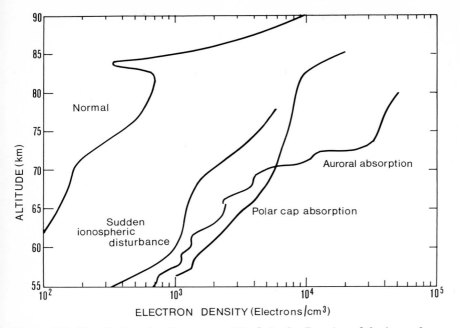

Figure 3.5. The electron density versus altitude in the D region of the ionosphere. Curves are drawn for typical normal conditions, sudden ionospheric disturbances, polar cap absorption, and auroral absorption. Taken from Bourdeau,[14] 1965.

increases by 2 orders of magnitude under severe conditions and this causes the "blackouts" that are seen.

3.3 E REGION

Watanabe and Hinteregger[18] have calculated the ionization in the E region in detail. They pointed out that the sun's energy in UV radiation in the interval from 800 to 1027 A is 0.54 erg/cm²-sec while the sun's X-ray energy from 10 to 170 A is only 0.2 erg/cm²-sec. The photoionization rate for UV from 800 to 1027 A is greater than for X-rays of 10 to 170 A. (See Fig. 3.6a.)

The electrons are lost by dissociative recombination on O_2^+. Therefore, they conclude that the E region ionization is due to UV and not to X-rays.

$$O_2^+ + e \rightarrow O + O. \tag{3.16}$$

This reaction then tends to control the electron density at equilibrium in the lower E region.

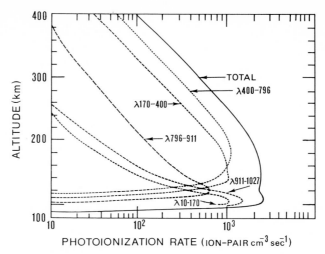

Figure 3.6a. Photoionization rates versus altitude at different wavelength intervals for vertical incidence. The total ionization rate is also included. Taken from Watanabe and Hinteregger,[18] 1962.

The photoionization rates versus altitude are plotted in Fig. 3.6a for different UV and X-ray wavelength intervals. The UV radiations from 170 to 1100 A have the highest photoionization rates in the E region. The ions O_2^+, O^+, and NO^+ are formed with the photoionization rates

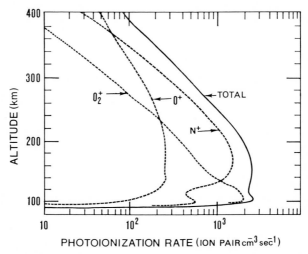

Figure 3.6b. Calculated photoionization rates for production of O_2^+, O^+, and N_2^+ versus altitude. Taken from Watanabe and Hinteregger,[18] 1962.

shown in Fig. 3.6b. The experimental data of Taylor and Brinton[19] for the percentage concentrations of these ions is shown in Fig. 3.6c. They measured the percentage compositions of the ion abundances with a Bennett radio frequency mass spectrometer on a rocket. N_2^+ was found to be a minor constituent while the NO^+ abundance was large. The NO^+ formation is usually explained by atom–ion interchange,

$$N_2 + O^+ \rightarrow NO^+ + N. \tag{3.17}$$

The ion compositions of Fig. 3.6c are in reasonable agreement with the calculations. In order to compute the semi-equilibrium values for the

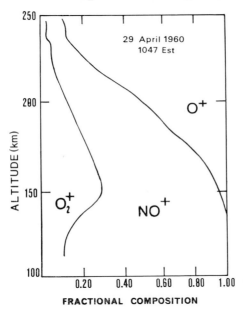

Figure 3.6c. Experimentally observed fractional ion concentration versus altitude for O_2^+, NO^+, and O^+ ions. Taken from Taylor and Brinton,[19] 1961.

electron intensity, dN_e/dt in Eq. 3.14 is set equal to 0. We use the value $q = 2000/cm^3$-sec at noon from Fig. 3.6b and obtain $N_e = 2.6 \times 10^5$ electrons/cm³ in agreement with rocket electron density measurements.

At night N_e drops below 10^3 electrons/cm³. It may be possible to maintain the nighttime E region by slow dissociative recombination. In that case it is not necessary to resort to an additional nighttime source.

Sporadic E ionization is sometimes seen at about 110 km altitude in

D

a thin layer of about 0.5 km thickness.   Electron densities as high as
10^5 electrons/cm³ are sometimes seen.   This is enough to noticeably
affect the high frequency radio wave refraction back to earth.

Densities of electrons in the E region are measured with ground based
ionosondes, radio wave propagation experiments, and plasma probes on
rockets.   One technique that was successfully used by Seddon and
Jackson[20] of the Naval Research Laboratory from 1946 to 1956 and by
Gringauz and associates [21] from 1954 to 1958 is rocket-to-ground radio
wave propagation.   The experiment works as a dispersion interferometer for radio waves.   Two frequencies, the fundamental and the
sixth harmonic, experience doppler shifts due to the rocket velocity

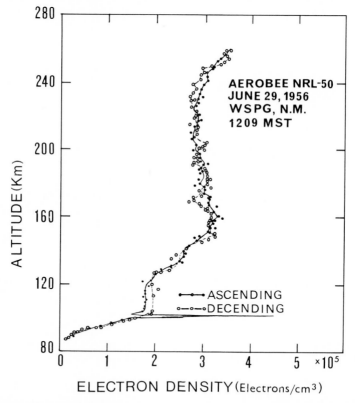

Figure 3.7. Electron density versus altitude for a rocket flight on June 29, 1956.
The open circles and the dashed line are plotted for the ascending leg and the solid
circles and the solid line for the descending leg.   A sharp spike of increased
electron density in the sporadic E layer was observed on the descending path.
Taken from Seddon and Jackson,[20] 1958.

The high frequency is unaffected by the ionosphere; the lower frequency has a higher phase velocity and thus a smaller index of refraction inside the ionosphere than outside. From the phase shift between the two signals the index of refraction is obtained and the electron density calculated.

An example[20] of the measured density in the E region up to 260 km is given in Fig. 3.7. The densities were obtained on June 29, 1956 during the time of the increase in the solar cycle. The sporadic E was seen on the descending leg at 100 km and would have been observed on the ascending leg, but good data then was not available.

Gringauz[21] was able to locate the peak of ionization in the F region with a rocket that went up to 473 km on February 21, 1958. That peak of 2×10^6 electrons/cm^3 occurred at an altitude of 280 km. It is this maximum in the F region that we now discuss.

3.4 F REGION

3.4.1 Electron Density

a. Ionosonde Measurements

Following World War II ionospheric stations all over the world were equipped with ionosondes that were used to measure the electron density versus height by the pulsed variable frequency method[7] described in the Introduction. The research concentrated on the study of the electron density anomalies in the F_2 region. These were not well behaved like the electron densities in the E region and in the F region where the F_1 layer was located. In one study in 1953 C. W. Allen[22] investigated the day–night variations of the F_2 critical frequencies and heights as a function of the time in the sunspot cycle, of the season of the year and of geographic and geomagnetic position. Data from 10 stations were analyzed. The chief characteristics of the F_2 critical frequencies could be described in terms of three anomalies; the sunspot minimum anomaly, the sunrise anomaly, and the day–night anomaly. During sunspot minimum the anomalies tended to be opposite in summer and in winter. In summer the critical frequencies had maxima at times near 2000 hr, rapid nighttime declines and deep presunrise minima. In winter the critical frequencies had minima near 2000 hr, rose during the night and had presunrise maxima. He concluded, "From the discussion it can be seen that ionospheric observations are not yet amenable to consistent explanation. The results of the present paper do not appear to improve the situation, but they are too regular and world-wide to be neglected."

Munro[23] used a network of five stations in the neighborhood of the University of Sydney at distances of about 50 km apart to study the travelling disturbances in the ionosphere. He found that the disturbances travelled horizontally at 5 to 10 km/min and vertically at about half the horizontal rate. Appleton,[24] also in 1958, found a daily variation in the abnormally low N_e values in the trough centered on the magnetic equator. The trough started to disappear during the after-

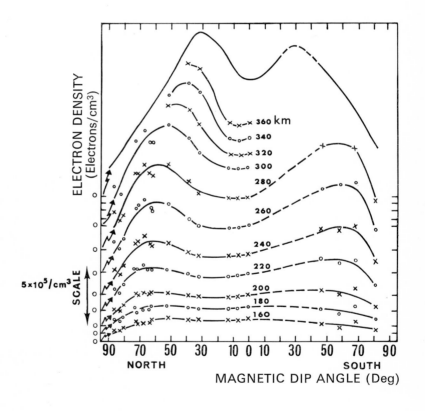

Figure 3.8. The electron density versus the magnetic dip angle at various altitudes for the F region at local noon in September 1957. The 0 of the graph for each curve is indicated on the side of the figure. This scale factor is determined by the arrows at the left of the figure. The average peak electron density N_{F_2} is given by the top curve. Taken from Croom, Robbins, and Thomas,[25] 1959.

noon and was replaced by a crest of abnormally high N_e values around 2100 hr. After midnight the equatorial crest fell rapidly and by 0300 hr had largely disappeared. The equatorial trough again developed in the late morning hours.

The equatorial minimum was studied[25] as a function of height with data from 18 stations using ionosondes. The results are given in Fig. 3.8 where the electron density is plotted against the magnetic dip angle for a number of different altitudes. The data were selected at local noon at equinox at the maximum of the sunspot cycle. The equatorial minimum occurred at all heights. At lower altitudes the dip was less and extended over a wider latitude interval. Similar curves were obtained for 0900 and 1500 hr local time. Another study[26] showed that the two maxima first developed at low latitudes and then shifted poleward during the day. The peaks reversed direction in the evening and then headed back toward the equator.

b. *Impedance Probes*

A radio frequency impedance probe experiment[27] was carried on the satellite Ariel. Data were recorded on a tape recorder for latitudes between 54 deg north and 54 deg south and for altitudes from 400 to 1200 km. The ionization was especially high on the magnetic shell of $L = 1.27$ after the Starfish nuclear bomb explosion over Johnson Island at 400 km altitude on July 9, 1962. That was the L value of the peak of the Starfish injected trapped electrons. The data also showed peaks of ionization on a lower shell at $L = 1.09$ and on a higher shell at $L = 1.75$. These last two shells of increased ionization were not explained.

c. *Topside Sounder*

A new method of investigating the ionosphere was introduced with the topside sounder on the spacecraft Alouette. It is an ionosonde that transmits pulsed radio waves from above the maximum in the iono- sphere rather than from below as for the ground based bottom side ionosonde. A transmitter and a receiver are on board so that both the density of electrons and the distance of the measured region below the satellite are determined. Many radio stations around the world received the telemetered data from the spacecraft and several groups analyzed the data. The early measurements[28] showed a trough at the equator with double humps that were farther from the equator at lower altitudes.

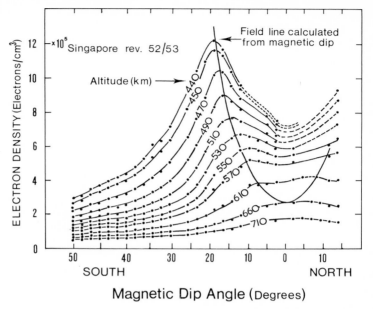

Figure 3.9. Electron density versus magnetic dip at various altitudes above Singapore at 1000 hr local time on October 1, 1962. The solid line is the magnetic field as calculated from the magnetic dip angles. The ionization is maximum along the magnetic field line shown. Taken from King, et al,[29] 1964.

At the higher altitudes, 650 km over Singapore, the humps merged into one peak. The humps were definitely aligned along a magnetic field line as shown in Fig. 3.9[29] where the electron density is plotted versus the magnetic dip angle. Note that the altitudes are reversed with the lowest altitude at the highest density. The magnetic field line is also upside down. These experimenters found that one peak only was present at 0900 hr in the morning, and that the peak split into two peaks that moved away from the equator to about 20 deg geomagnetic latitude by 1400 hr. At the higher latitudes Fig. 3.10, a plot of the critical frequency versus local time, shows that the maximum occurs latest in the day at the highest geomagnetic latitudes.[30] The humps of the F region appear to move toward the magnetic equator during the morning from 0300 to 1500 hr and away from the equator during the afternoon and night from 1500 to 0300 hr.

Many interesting frequency traces are observed on the frequency versus height ionograms from the topside sounder.[31] A photograph of one obtained at the Stanford satellite monitoring facility at 2140 hr on

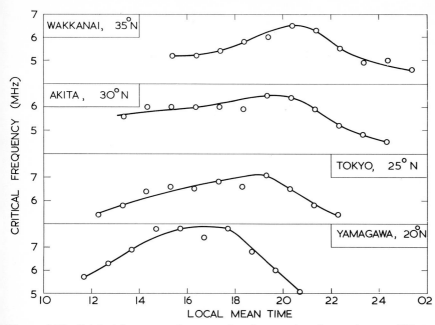

Figure 3.10. Critical frequency f_{F_2} versus local mean time for stations at different magnetic latitudes. The value of the magnetic latitude appears by each curve. The zeros for each curve appear on the left side of the graph. The data are the monthly medians for June 1963. The maximum occurs later and later as the magnetic latitude increases. Taken from King, *et al.*,[30] 1964.

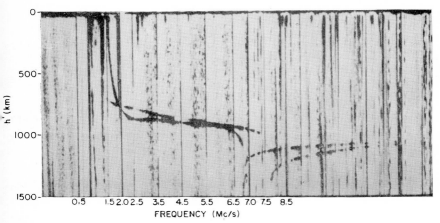

Figure 3.11a. A photograph of a topside ionogram obtained at the Stanford monitoring facility at 2140 hr geomagnetic time on March 15, 1963. The plasma resonance and cyclotron harmonic spikes are clearly visible. Taken from Thomas and Sader,[31] 1964.

March 15, 1963 is shown in Fig. 3.11a. In the picture the plasma and the cyclotron resonance spikes point downward from the top and the ordinary and extraordinary ray traces are sketched for easier identification in Fig. 3.11b.

For electromagnetic waves to propagate the index of refraction of the plasma must be real and finite. Certain magnetoionic conditions are then required for propagation. These conditions and the defining equations are given in Table 3.1 and the definitions of the symbols in Table 3.2.

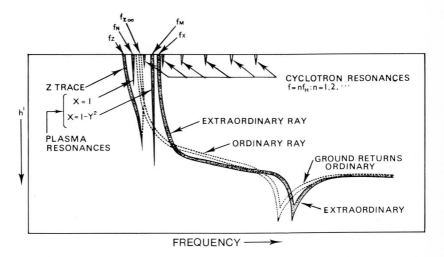

Figure 3.11b. A sketch of a topside ionogram which identifies the various radio signals. The magnetoionic conditions required for propagation and the defining equations are given in Table 3.1 and the definitions in Table 3.2. Taken from Thomas and Sader,[31] 1964.

The extraordinary ray f_x is usually the clearest on the ionogram and is easily recognized at small distances h' from the spacecraft. From Eqs. (3.5) and (3.24) we find

$$N_e = (1.24 \times 10^{-8})\,(f_x^2 - f_x f_{Be})\ \text{electrons/cm}^3. \qquad (3.18)$$

The cyclotron harmonics and Eq. (3.19) may be used to obtain the magnetic field at the satellite. And the value of the dip angle at the satellite θ can be found from Eq. (3.22).

Table 3.1 Relevant magnetoionic conditions

In this table $X = f_N^2/f_0^2$, $Y = f_{Be}/f$. Taken from Thomas and Sader.[31]

Magnetoionic condition	Remarks	Frequency satisfying magnetoionic condition	Equation	
$Y = 1/n$, $\quad n = 1, 2, \cdots$	Cyclotron harmonics	nf_{Be}	$f = nf_{Be}$	(3.19)
$X = 1 + Y$	Z trace zero range echoes	f_z	$f^2 = f_N^2 - f_z f_{Be}$	(3.20)
$X = 1$	Ordinary trace zero range echoes	f_0	$f_0^2 = f_N^2$	(3.21)
$X = 1$	Plasma resonance	f_N $\Big\}$		
$X = \dfrac{1 - Y^2}{1 - Y^2 \cos^2(\pi/2 - \theta)}$	Z trace infinite range echoes	$f_{z\infty}$		(3.22)
$X = 1 - Y^2$	Magnetic plasma resonance	f_M	$f_M^2 = f_N^2 + f_{Be}^2$	(3.23)
$X = 1 - Y$	Extraordinary trace zero range echoes	f_x	$f_x^2 = f_N^2 + f_x f_{Be}$	(3.24)

Table 3.2. Definitions

Frequency.

f_x Frequency at which ordinary ray has zero range ($X = 1$). This is the same as the plasma frequency at the vehicle f_N.

f_z Frequency at which extraordinary ray has zero range ($X = 1 - Y$).

f_M Frequency at which Z ray has zero range ($X = 1 + Y$).

Plasma resonance ($X = 1 - Y^2$).

f_N Plasma resonance ($X = 1$). This is the plasma frequency at the vehicle and is the same as the frequency for the ordinary trace zero range echoes.

$f_{z\infty}$ Frequency at which Z trace has infinite virtual depth

$$\left(X = \frac{1 - Y^2}{1 - Y^2 \cos^2(\pi/2 - \theta)} \right).$$

f_{Be} Electron gyrofrequency.

N_e Electron density at the vehicle.

θ Dip angle at the vehicle.

X f_N^2/f^2.

Y f_{Be}/f.

h' Distance from spacecraft down to effective reflection level.

3.4.2 Electron temperature

The electron temperature and density were both measured with Langmuir type electrostatic probes on Explorer 17. A block diagram of the experiment[32] is given in Fig. 3.12. A sawtooth voltage of 0 to $+0.75$ V was applied to the collector that protruded through a guard in the satellite skin and the detector current was read.

Several general characteristics of the ionosphere were observed: (1) a steep morning rise in the electron temperature T_e with a morning maximum about five hours after local sunrise; (2) a decrease in T_e to an afternoon plateau that declines steeply at sunset; (3) a nighttime plateau of T_e at about 1150 deg that is moderately variable. The authors concluded that the electron temperature was always greater

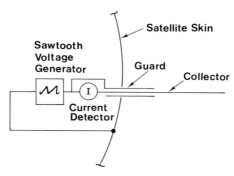

Figure 3.12. Functional block diagram of the electrostatic probe experiment. A sawtooth voltage applied to the guard and collector produces a volt–ampere characteristic from which either T_e or N_i can be derived. Taken from Brace and Spencer,[32] 1964.

than the neutral molecule temperature both day and night. The difference in temperatures was always greater at high than at low altitudes so a local heat source at night of 15 eV/cm^2-sec was needed. Since the sun does not shine at night another source was required. Finally N_e and T_e versus latitude changed in opposite directions. When N_e was large, T_e was small and vice versa.

The temperature variations with latitude and with solar storms were investigated with Langmuir probes on the Ariel satellite.[33] The temperature was invariably higher at high than at low latitudes by as much as 150 to 700° K. The dawn increase in temperature was strongest at the magnetic equator and was explained[34] by the morning lag of the increase in electron density. During this time the collisional cooling rate of the electrons was low relative to the heating rate. A strong negative correlation between density and temperature was observed at all latitudes.

The variation of electron temperature with UV light from the sun was investigated using the 2800 mc/s emission as a monitor. The values for the temperature changes were (4.6 ± 2.3) °K/flux unit during the day, and (-2.3 ± 2.7) °K/flux unit at night. One flux unit is 10^{-22} W/m²–c/s.

During magnetic storms it was found that the electron density invariably increased and that the temperature usually decreased. Furthermore, the increase of the electron density was simultaneous in time with the decrease of electron temperature. This strong negative correlation of density and temperature is similar to the quiet time anti-correlation described above. As before, the decreased temperature is attributed to the increased collision cooling of the electrons.

The latest temperature measurements have been made by Brace, Reddy, and Mayr[35] with cylindrical electrostatic probes on Explorer 22. They concluded that: (1) an equilibrium exists between the electrons at 1,000 km and the electrons and H and He atoms at higher altitudes, (2) the latitude distributions of N_e and T_e near the equator are consistent with this equilibrium day and night; and (3) local electron cooling exists in the equatorial region. At 1,000 km a H density of 5×10^5 atoms/cm³ is required. This is 10 times higher than the H used in most atmospheric models.

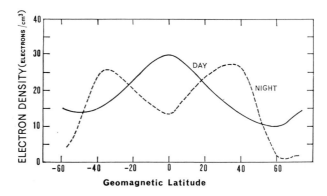

Figure 3.13a and b. Electron density and temperature versus geomagnetic latitude. At 1,000 km during the 1965 vernal equinox the density and temperature during the day is shown by the solid line and during the night by the dashed line. Taken from Brace, Reddy, and Mayr,[35] 1967.

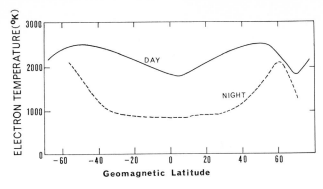

Figure 3.13b.

The day and night differences of N_e and T_e versus geomagnetic latitude are shown in Fig. 3.13a and 3.13b. The solid line is plotted for the day and the dashed line for the night in the Spring of 1965 at 1,000 km. The peak electron density at the equator in the day turns into a minimum during the night. The temperature, on the other hand, always has a minimum at the equator, a maximum at 60 deg geomagnetic latitude, and is higher in the day than at night. The inverse relation between N_e and T_e is seen by comparing the solid curves for N_e and T_e in Fig. 3.13a and 3.13b.

A summary of the experimental values for N_e and T_e is given in Fig. 3.14a and 3.14b. The electron density and electron temperature are plotted vertically and the geomagnetic latitude, horizontally. Each curve represents a different time of day. The continuous evolution of N_e from a peak during the day to a valley during the night is clearly shown. Likewise, the temperature trough at the equator, which varies through the day, and the maxima at 60 deg geomagnetic latitude are evident in Fig. 3.14b. The important processes controlling the 1,000 km ionosphere are the photoelectron heat source and the electron cooling by the local electrons and protons. The equatorial maximum of N_e is explained[36] as the diffusion upward of electrons and ions along magnetic field lines in response to the daytime heating by photoelectrons. The equatorial minimum of N_e at night is caused by the downward diffusion of electrons and protons because of nighttime cooling. Because the magnetic field lines at the low latitudes are shorter, the total number of electrons is correspondingly smaller and the total heat content less. More day–night variation is expected because a larger fraction of heat

ELECTRON DENSITY

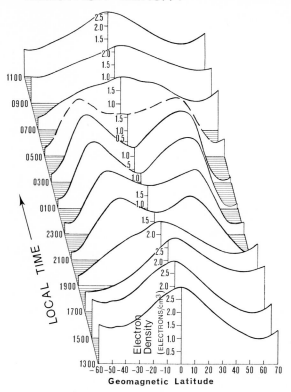

Figure 3.14a. Average latitude and local time variation of the electron density at 1000 km at the 1965 vernal equinox versus geomagnetic latitude. The development of the peak at the geomagnetic equator during the morning and afternoon and the appearance of a trough at night is clearly shown. The dashed line indicates where the data is interpolated. Taken from Brace, Reddy, and Mayr,[35] 1967.

drains away at night. At the higher latitudes, magnetic field lines are longer and the heat content is greater. The photoelectrons interact on their paths to the opposite conjugate points and deposit nearly all their energy to heat the high ionosphere. The large heat sink during the day changes to a large heat source at night and holds the temperature nearly constant at high latitudes.

The authors[35] conclude, "(1) The equatorial protonosphere near 1,000 km is always in diffusive equilibrium, even during the periods of the

ELECTRON TEMPERATURE

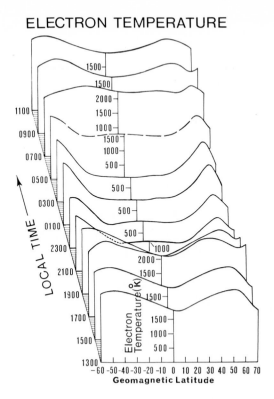

Figure 3.14b. Average latitude and local time variation of T_e at 1,000 km at the 1965 vernal equinox versus geomagnetic latitude. The trough at the equator is always present but varies in height during the day. The dashed line indicates where data is interpolated. Taken from Brace, Reddy, and Mayr,[35] 1967.

most dynamic change at sunrise and sunset. (2) The daytime mimimum of T_e observed at the equator is caused by local cooling via protons to an unexpectedly great amount of neutral hydrogen (about $4 \times 10^5/\text{cm}^3$). (3) The observed daytime equatorial maximum of N_e is consistent with the T_e minimum and is enhanced by the predominance of H^+. (4) The flat nighttime trough of T_e is quantitatively in agreement with the steep trough of N_e found at the equator. (5) The sunset decay of T_e and N_e at the equator requires a local cooling that is also consistent with a high concentration of neutral hydrogen ($\approx 5 \times 10^5/\text{cm}^3$)." There is every reason to believe that these processes are also important at other altitudes.

3.4.3 Ion Composition and Density

a. *Retarding Potential Probes*

The ion density was measured[37] by an ion probe that was boosted on a Javelin rocket to an altitude of 1,000 km over Wallops Island at 2044 hr local time on November 9, 1960. The ion density was measured from 240 to 700 km. A sketch of the ion trap is given in Fig. 3.15a. The first grid is held at the rocket skin ground. The second grid is swept from − 2.4 V to + 7.0 V in 2 sec. The third grid was held at − 58 V to repel the secondary electrons from the collector fixed at − 30 V. The ion densities measured with ion traps are plotted as circles in Fig. 3.15b and compared to ionosonde data plotted as triangles. Where the two sets of data overlap the agreement was very good.

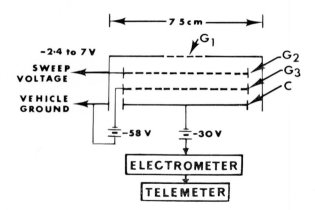

Figure 3.15a. A schematic drawing of an ion-trap. Grid 1 is at vehicle ground potential. Grid 2 is swept from − 2.4 to 7 volts, grid 3 is − 58 volts, and grid 4 is − 30 volts. Taken from Hanson and McKibbin,[37] 1961.

The first experimental evidence[38] for the presence of helium ions was obtained from a plasma probe on the satellite Explorer 8 that was launched on November 3, 1960 into an orbit with 50 deg inclination. The satellite carried four plasma probes. Two measured electron temperature, one the positive-ion concentration and the other the mass distribution of positive ions as a function of the applied retarding potential when the detector was pointed in the direction of the satellite motion. A photograph of the ionospheric satellite is shown in Fig. 3.16a. The arrows point to the instruments: (1) retarding-potential

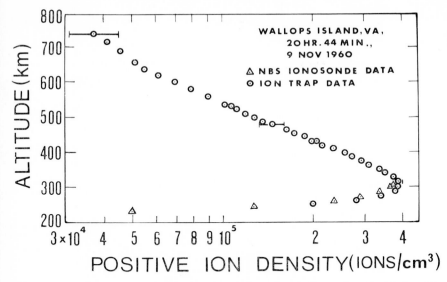

Figure 3.15b. Positive ion density versus altitude on November 9, 1960. The triangles are ionosonde data and the circles are ion trap data. Taken from Hanson and McKibbin,[37] 1961.

probe, (2) ion-current monitor, (3) electron-temperature probe, and (4) aspect sensor.

As the voltage on the retarding-potential probe is increased, the ions with kinetic energy less than the applied voltage are repelled. The kinetic energy of the ions relative to the spacecraft is

$$E = \tfrac{1}{2}\, m_i v_s^2, \tag{3.25}$$

where m_i is the mass of the ion and v_s is the satellite velocity.

For a spacecraft velocity of 6 km/sec the kinetic energies for He$^+$ and O$^+$ are 0.75 and 3.0 eV. Since a temperature of 1,000° K corresponds to 0.09 eV, the temperature will only broaden the He$^+$ energy line. For reasonable temperatures and good resolution, downward steps are observed in the ion current as the retarding voltage is increased. At the lowest voltages all ions can enter the detector. As the voltage is increased first the light and then the heavier ions are repelled. Figure 3.16b gives the ion current normalized to one at 0 volts. The circles are the experimental values and the solid lines numbered 1 and 2 are the He$^+$/O$^+$ ratios of 1.0 and 1.5. The data were taken at 1115 UT (universal time) on November 24, 1960 at an altitude of 1630 km at

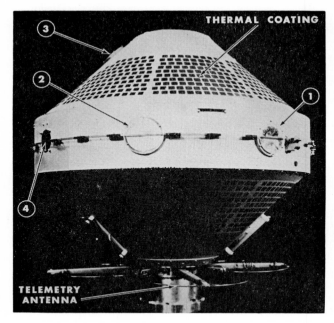

Figure 3.16a. The Explorer VIII satellite showing (1) retarding probe, (2) ion current monitor, (3) electron temperature probe, and (4) aspect sensor. Taken from Bourdeau, *et al.*,[38] 1962.

75 deg west longitude and 32 deg south latitude. The data gave about equal amounts of He^+ and O^+ ions at this time and place.

The H^+ abundance as well as the He^+ and O^+ abundances were measured[39] with a spherical retarding-potential probe on the satellite Ariel I that was launched in April 1962. A sawtooth voltage of -3 to 20 V once a minute was applied to the 9 cm diameter collector sphere. The sphere was surrounded by a 10 cm grid at -6 V with respect to the satellite skin to repel the electrons. The results are shown for July 1962 in Fig. 3.17a where the probe current is plotted against a number proportional to the probe voltage. The steps are labelled where the H^+, then He^+, and finally O^+ ions are repelled as the voltage increases. There is no evidence for N^+. An indication of how the composition changes with latitude is given in Fig. 3.17b. Contours of constant percentage composition of O^+ in an O^+–He^+ mixture are indicated for an altitude of 1,000 km on the latitude–longitude map. The solid dots indicate an O^+ abundance of 10 per cent, the squares 25 per cent, the

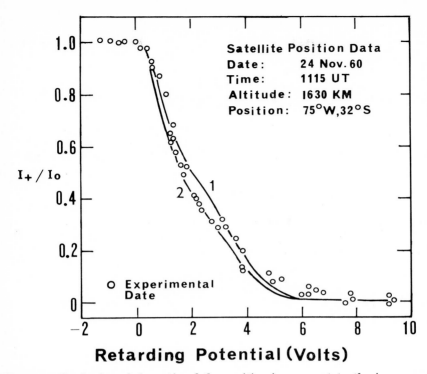

Figure 3.16b. A plot of the ratio of the positive ion current to the ion current at a retarding potential of 10 volts versus retarding potential. The circles are the experimental data, the solid lines numbered 1 and 2 are the calculated He$^+$/O$^+$ ratios of 1.0 and 1.5. Other calculated He$^+$/O$^+$ ratios are much farther from the data. Taken from Bourdeau, *et al.*,[38] 1962.

triangles 50 per cent, and the circles 80 per cent. The larger abundance of the He$^+$ at the equator and the heavier O$^+$ at higher latitudes is clear. The day–night variation is very marked. At night the He$^+$ might be the predominant ion at 600 km and H$^+$ at 1,200 km but in the early afternoon O$^+$ predominates up to 1,200 km and the transition to H$^+$ is at an altitude much higher than the satellite.

b. Lower Hybrid Resonance

The combination of topside sounding and the very low frequency noise experiment on Alouette led to the discovery of a new unexpected signal, the lower hybrid resonance that is useful in determining the mean effective mass of the ionospheric ion, $\overline{m}_i$.[40] The lower hybrid

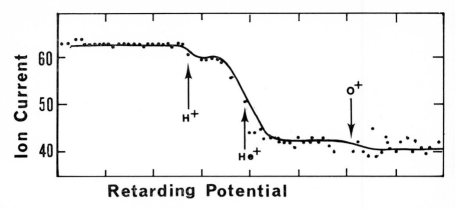

Figure 3.17a. Ion current versus retarding potential. The data points are represented by black dots. The solid curve is a best fit to the data. The positions where H+, He+, and O+ can no longer enter the detector are indicated by arrows. Data were taken at 1,000 km at 53 deg north latitude on May 11, 1962. Taken from Bowen, *et al.*,[39] 1964.

resonance, LHR, is triggered by whistlers that interact with the plasma in the immediate vicinity of the satellite. The LHR signals were discovered accidentally when the low frequency radio receiver was left on simultaneously with the topside sounder. These LHR resonances were seen in a large fraction of the cases when the radio receiver was left on. The LHR are never observed as VLF noise at ground receivers. The equation for the hybrid frequency is

$$f_{LHR} = \left[\frac{f_N^2 f_{Be}^2}{f_N^2 + f_{Be}^2} \right]^{1/2} \left[\frac{m_e}{m_p} \frac{1}{\overline{m}_i} \right]^{1/2}. \tag{3.26}$$

The $\overline{m}_i$ is the mean mass number of the positive ions in the medium and is defined by

$$\frac{1}{\overline{m}_i} = \sum_i \frac{m_p}{m_i} \frac{N_i}{N_T}. \tag{3.27}$$

The m_p is the proton mass, N_i is the number density of the ith ion, and N_T is the total ion or electron number density. Since f_{LHF} and f_N are measured and f_{Be} can be calculated or measured, the value of N_i is readily obtained. The values for $\overline{m}_i$ vary from close to one at the equator to about nine at $L = 10$ at 1,000 km altitude. Limits on the concentrations of the constituents H+, He+, and O+ can be derived from the values of $\overline{m}_i$, i.e., H+ varies from 60 per cent at $L = 1.3$ to less than

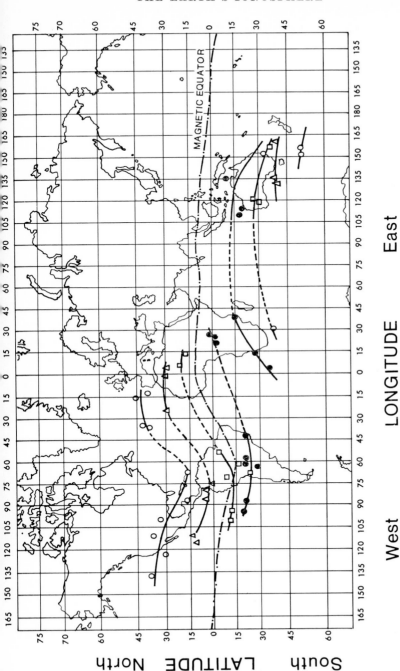

Figure 3.17b. Ion composition contours on a latitude–longitude map corrected for height to 1,000 km. The percentage of O^+ in an O^+/He^+ mixture is given. The open circles indicate 80 per cent, triangles 50 per cent, squares 25 per cent, and solid circles 10 per cent. The contours follow the geomagnetic latitudes which show the large amount of magnetic field control. Taken from Bowen, *et al.*,[39] 1964.

about 10 per cent at $L = 5$. On the other hand, O^+ varies from less than 30 per cent at $L = 1.3$ to between 50 and 80 per cent at the highest L values. Limits on the maximum and minimum effective temperatures can be calculated from the effective ion mass and the scale height of the electrons. In the auroral and polar regions, the temperatures are sometimes as high as 3,000 to 4,000° K.

3.4.4 Ionospheric disturbance correlations

a. With Trapped Radiation

A correlation[41] between the critical frequency f_{F_2} and the fluxes of electrons in the outer radiation belt was found for a minor magnetic disturbance at the end of September 1962. Both decreased, then took several days to return to their initial values. Therefore, the ionization in the F region and the trapped electrons decreased and increased together. On a pass of Alouette on October 24, 1962 an unexplained depression in the electron density at 45 deg north latitude was found. A plot of the contours of constant f_{F_2} on a height versus geomagnetic latitude plot is shown in Fig. 3.18. The 45 deg depression is at the same position as the outer electron radiation belt. Figure 3.18 also shows that the contours of critical frequency have their maxima over the magnetic equator at altitudes greater than 700 km. At lower altitudes the equatorial trough is evident in agreement with Figs. 2, 8, and 9.

b. With Precipitated Electron Flux

At times of ionospheric disturbances the ion and electron densities in the D and E regions usually increase. The absorption of radio waves increases. Since the absorption varies inversely as f^2, the minimum observed frequency $f_{\min}$ increases sometimes until all frequencies are absorbed. The result is a total blackout. Usually the $F2$ layer critical frequency f_{F_2} decreases and its virtual height h'_{F_2} increases.

A correlation[42] has been shown between the ionospheric disturbances and the electron flux lost from the outer radiation belt, the precipitated electron flux "p.e.f.", near the magnetic shell $L = 4$. The ionosphere was classified as "disturbed" or "quiet" on the basis of the values of f_{F_2} and h'_{F_2} for the 10 magnetically quiet days of each month. The

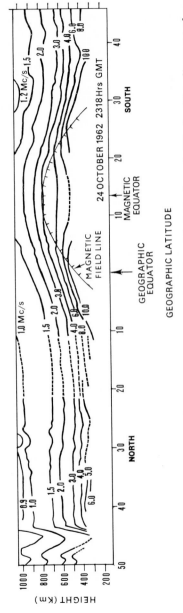

Figure 3.18. Contours of constant critical frequency f_{F_2} versus geographic latitude and altitude. The alignment along the magnetic field line on both sides of the magnetic equator is shown. A sharp dip at 45 deg north latitude is also present. Taken from Nelms,[41] 1963.

ionosphere was designated as disturbed if either f_{F2} or h'_{F2} fell outside the quiet range within one hour of the p.e.f. measured by the satellite detector at the point or at its conjugate point. The satellite data were always read out in the northern hemisphere so only conjugate p.e.f. data were available for comparison with the magnetic disturbances over the ionsopheric stations in the southern hemisphere.

The six ionospheric stations used in the study and their latitudes, longitudes, L values, critical p.e.f. values, and critical p.e.f. values for the conjugate points for the southern stations are given in Table 3.3. A positive correlation is counted if the observed p.e.f. is greater than the critical one. The results of the study are given in the last two columns. The percentage of the time that the p.e.f. is greater than the critical value is given in the next to the last column and the percentage of the time that the ionospheric disturbances occur in the last column. The agreement between the last two columns at all stations was extremely good. Another way of expressing the good correlation is that a total of 88 agreements between disturbed conditions and high precipitating electron fluxes were seen and zero cases of quiet conditions and high fluxes.

c. With Visible Aurora

It is now possible to continuously monitor the visible aurora by radio methods. A comparison[43] of simultaneous optical and radio auroral measurements during 12 days in December, 1964 showed conclusively that the radio wave echoes originate in the ionization of the visible aurora. The College Alaska multi-frequency backscatter radio data were compared with the optical data from the Alaskan network of all-sky cameras located between 65 and 80 deg north geomagnetic latitude. The radio slant range of each signal was converted directly into ground distance. The range width of the signals gave the latitude span of the scattering region.

The positions of the aurora as determined by radio and optical methods agreed well except when the aurora was nearly overhead. For that reason a future aurora radio monitor station should be located somewhat south of College. The radio method has the advantage that it can be used to monitor the aurora during the four summer months when no optical measurements are possible. Because radio measurements will be independent of the weather, studies of the seasonal and day–

Table 3.3 Frequency of p.e.f. exceeding critical values and of ionospheric disturbances at six stations near $L = 4$

Station	Latitude	Longitude	L	Critical p.e.f. in conjugate area, electrons/cm²-sec	Critical p.e.f. at station, electrons/cm²-sec	% Time p.e.f. > critical value	% Ionospheric disturbance
SANAE	70.3° S	2.4° W	3.99	8.2×10^3	2.5×10^4	67	65
Campbell Island	52.55° S	169.15° E	4.02	2.5×10^4	2.2×10^4	29	31
Halley Bay	75.52° S	26.60° W	4.17	1.2×10^4	2.5×10^4	53	54
St. Johns	47.52° N	52.78° W	3.47	—	2.5×10^4	24	23
Ottawa	47.40° N	75.75° W	3.64	—	2.2×10^4	28	26
Winnipeg	49.90° N	97.40° W	4.22	—	2.1×10^4	30	31

night variations in the aurora can be carried out that were not possible in the past.

Perhaps all ionospheric disturbances can be explained by ionization caused by charged particles stopping in the atmosphere. Those occurring under the radiation belts are caused by electrons precipitating out of the belts. Those occurring in the aurora and polar regions may be caused by precipitating auroral and magnetic cavity tail electrons and protons. At times of solar proton flares, fast protons bombard the polar regions and the resulting ionization causes the polar cap absorption of galactic radio noise. The increased solar wind plasma that arrives two days later causes the magnetic storms. One known exception to the charged particle induced ionization is that caused by the X-rays in the D region during a solar flare.

3.5 HIGH IONOSPHERE

3.5.1 Electron density

a. Whistlers

The electromagnetic radiation from lightning discharges travels upward along the magnetic field lines and is observed at the opposite or at the original ends after reflection. These waves are called whistlers. A photograph of a frequency time trace of a whistler recorded at Eights, Antarctica is shown in Fig. 3.19. The travel times of the different frequencies depend on the electron densities encountered. The dispersion of the frequencies is used to measure the electron density. The density at the outermost extent of the line at the equator is most effective and therefore is the one measured.

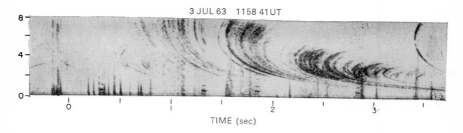

Figure 3.19. Photograph of the frequency versus time trace of a whistler recorded at Eights, Antarctica, 64 deg south latitude. Nose whistlers with different nose frequencies versus time are present. Taken from Carpenter and Smith,[47] 1964.

The whistlers travel in columns or ducts of increased ionization aligned along the magnetic field lines. The whistler energy is trapped inside these ducts by internal reflection in the same way as in light pipes.

High atmosphere studies were intensified after the discovery[44] of the 'nose" whistler. In these whistlers the frequencies both increase and decrease with time from the beginning called the nose. Trains of nose whistlers are made up of whistlers that follow different magnetic field lines. The nose frequency gradually decreases with time. When the plasma frequency f_N, cyclotron frequency f_{Be}, and wave frequency f are of magnitude such that f_N is much greater than f and f_N is about equal to or greater than f_{Be} then the group velocity is

$$v_g \simeq \frac{2cf^{1/2}(f_{Be}-f)^{3/2}}{f_N f_{Be}}. \tag{3.28}$$

The nose frequency is the frequency for which v_g is a maximum, $f_{nose} = f_{Be}/4$. At frequencies $f \ll f_{nose}$, v_g depends on B and N_e as $v_g \sim \sqrt{B/N_e}$ and the time delay for a path length s is

$$\Delta t = s/v_g \sim \sqrt{N_e/B} \ s. \tag{3.29}$$

Whistlers are clearly useful for investigating the electron density and magnetic field at great distances from the earth.

Smith[45] used the velocity of Eq. 3.28 to calculate the time delay Δt over the whistler path

$$\Delta t = \frac{1}{2c} \int_{path} \frac{f_N f_{Be}\, ds}{f^{1/2}(f_{Be}-f)^{3/2}}. \tag{3.30}$$

The travel time becomes a maximum when f approaches zero and when the cyclotron frequency is a minimum which occurs at the highest latitudes. In between, the minimum time occurs at the nose frequency. The dispersion is given by

$$D = \Delta t f^{1/2} = \frac{1}{2c} \int_{path} \frac{f_N f_{Be}}{(f_{Be}-f)^{3/2}}\, ds. \tag{3.31}$$

For $f \ll f_{Be}$, the dispersion approaches the constant defined previously in Eq. 3.9.

From Eq. 3.29 we see that the outer ionosphere density is proportional to B. For a dipole field B is proportional to $1/L^3$ at the equator and then N_e is given[45] by

$$N_e \text{ (electrons/cm}^3) = 1.2 \times 10^4 f_{Be} \text{ (c/s)}. \tag{3.32}$$

b. Knee Whistlers

At several earth radii the electron density derived from whistlers decreases abruptly. The rapid decrease in density is called a "knee", and the corresponding whistler a "knee whistler". Carpenter[46] compares the knee whistler with an ordinary whistler in Fig. 3.20. The frequency train and the nose frequency versus time, and the electron density versus distance are shown in the curves of Fig. 3.20 a, b, and c on the lefthand side for the ordinary nose whistler. Similar curves of Fig. 3.20 d, e, and f are shown on the righthand side for the knee whistler. The knee whistler is distinguished by the sudden increase in nose frequencies indicated in Fig. 3.20 d and e and the resulting sudden decrease in electron density with distance shown in Fig. 3.20f.

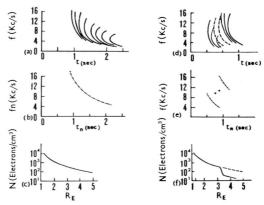

Figure 3.20a, b, c, d, e, and f. Wave frequency and nose frequency versus time and electron density versus distance sketches for ordinary nose whistlers in Fig. 3.20a, b, and c, and for knee whistlers in Figures 3.20d, e, and f. The curve in Figure 3.20c is an average of the results for December and June near sunspot maximum. Taken from Carpenter,[46] 1963.

Normally the transit time is given by Eq. 3.29. For an electron density that falls off as the earth's magnetic field, N_e/B is a constant. The Δt is then proportional to the path length along the magnetic field line which is longer at the higher latitudes. So the transit time is longer for greater distances from the earth. However, if N_e decreases abruptly N_e/B is no longer a constant, N_e decreases much faster than s increases, and Δt decreases with increasing distances from the earth. At even greater distances N_e decreases more slowly again and the train looks normal. This results, however, in superimposing the nose train for great distances over the one nearer the earth. The resulting electron density curve is sketched in Fig. 3.20f.

An excellent review[47] of whistlers to 1964 discusses nose whistlers, knee whistlers and models of the electron density in greater detail. The knee is thought to be permanent. It moves to lower L values at times of increasing magnetic storms. During severe storms it may be found below $L = 3.5$, during moderate storms between $L = 3$ and 4.5 and at quiet times between $L = 4.5$ and 6.

c. Variations in Density

The outer ionospheric electron densities are usually lower during the main phase and recovery periods of magnetic storms.[47] For $L < 4$ the density is often reduced by 20 per cent and in severe storms by as much as a factor of 10. In severe storms the electron density may be reduced by a factor of 2 at $L = 2$ in a time of four hours and by a factor of 10 at $L = 2$ to 3 in 6 to 36 hours.

Nose whistler data strongly suggest a seasonal variation of electron density with maxima in October–November and April–May. At Portier, France, 49.5 deg geomagnetic latitude, the ratio of minimum to maximum densities is 0.75. The yearly variation has a maximum in December and a minimum in June with an amplitude of about 30 per cent that decreases with decreasing solar activity.

From $L = 2.5$ to $L = 5$ the electron density change in the years 1958 to 1961 was only 20 per cent. The changes over the solar cycle are not expected to be great.

Day–night studies[48] of electron densities showed slow inward movement of the knee of 1.5 earth radii during the 10 hours at night, a slight shift outward of 0.5 earth radii during the day and a rapid outward shift of 1.0 earth radii in one hour in the late afternoon. This last movement is probably caused by the injection of new plasma of 100 electrons/cm³ at four to five earth radii. The electrons and ions inside the boundary knee appear to rotate with the angular velocity of the earth. Carpenter's[48] model of the ions and electrons in equatorial cross section is shown in Fig. 3.21. The bulge shows the regions of plasma injection on the evening side of the earth.

Outside the plasma boundary the electron density[49] drops to one electron/cm³ at seven earth radii. This is about a factor of four lower than in the region between the magnetic cavity boundary and the shock front formed by the solar wind incident on the earth's magnetic field.

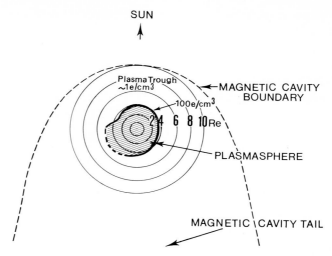

Figure 3.21. A model of the equatorial cross section of thermal ions in the magnetic cavity. The model is based on data for June through August 1963 during moderate steady magnetic field conditions, $K_p = 2$ to 4. The sun's direction, the boundary of the magnetic field cavity, the region of about 1 electron/cm³ outside the knee, and the region of 100 electrons/cm³ at the knee are given. The boundary of the plasma region showing the position of injection of new plasma on the west side is indicated. Taken from Carpenter,[48] 1966.

3.5.2 Positive ion density

Positive ions were measured[50] by ion traps on two Soviet spaceships launched on January 2 and September 12, 1959. Ion densities of about 10^3 ions/cm³ were found at distances out to 22,000 km. The density then fell rapidly to a value probably less than 30 ions/cm³. This measurement of the density knee actually preceded the whistler measurements and is in good agreement with them.

3.5.3 Positive ion composition

The positive ion composition from 1,500 to 30,000 km was measured[51] with two Bennett radio frequency mass spectrometers on the Ogo-A satellite. The low mass spectrometer had a resolution of 0.5 atomic mass units and the high mass spectrometer of one atomic mass unit. It covered a current sensitivity of 10^{-9} to 10^{-14} amp. To minimize the constant photoemission background a synchronous detection system analyzed the ac component of the detected ion beam that was modulated by a 155 c/s signal. A saw-tooth sweep produced the negative accelerating potential used in analyzing the ions.

H$^+$ and He$^+$ distributions were obtained over 15 orbits from September 23 to December 10, 1964. Inbound data were obtained on the night side of the earth and outbound data on the day side. Results for three passes are shown in Fig. 3.22a where the ion density plateaus are closer to the earth than $L = 3$ and three passes in Fig. 3.22b where they

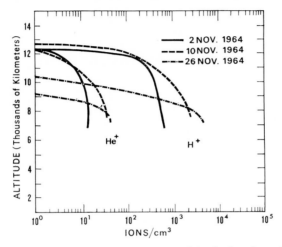

Figure 3.22a. H$^+$ and He$^+$ ion densities versus altitude for three daytime passes during November 1964. The data were taken with the Bennett radio frequency mass spectrometer on the Ogo-A satellite. Note the knees for H$^+$ and He$^+$ at around 10,000 km. Taken from Taylor, Brinton, and Smith,[51] 1965.

are farther away, $L = 4.5$ to 5.5. The H$^+$ ion density reaches a maximum of 10^3 ions/cm^3 at about 4,000 km and is constant out to the drop off at the plateau. There the density drops a factor of 10 or more in a distance at only 500 km. The He$^+$ ion density is consistently one per cent of the H$^+$ ion density over most of the altitudes.

The H$^+$ and He$^+$ ion density contours are very similar to the whistler electron density contours and to the ion trap measurements. This is even more significant because the electron densities were measured at the equator and the ion densities at a variety of latitudes. The region of thermal ions appears to expand and contract with changes in magnetic activity. The distribution of ions appears to be controlled by the geomagnetic field. The distance to the sharp decrease in ion density varies inversely with magnetic activity.

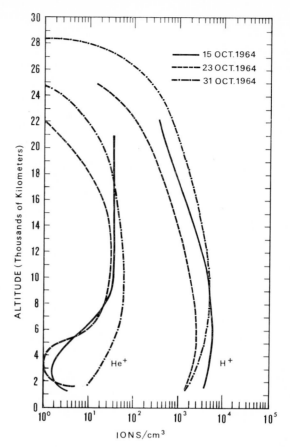

Figure 3.22b. H+ and He+ ion densities versus altitude from 1,000 to 30,000 km. These data were taken for three nighttime passes in October 1964. They show that ion densities increase to maxima at a few thousand km and then fall off at the knee at about 25,000 km. The H+/He+ ratio is always about 100. Taken from Taylor, Brinton, and Smith,[51] 1965.

3.5.4. Positive ion temperature

A new kind of radio signal, the proton whistler,[52] was identified from recordings by Injun 3 and Alouette satellites. It is the left-hand-polarized ion cyclotron wave. A photograph and sketch of the proton whistler is shown in Fig. 3.23a and b. It is preceeded by the ordinary right-hand whistler, the electron whistler. The cross-over frequency ω_{12} is the frequency at which both electron and proton whistlers can

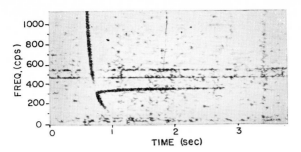

Figure 3.23a. A photograph of a frequency versus time sonogram of an electron and a proton whistler. Taken from Gurnett, *et al.*,[52] 1965.

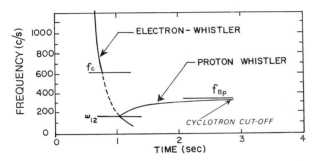

Figure 3.23b. A sketch of the sonogram showing the proton whistler, electron whistler, and associated frequencies. f_{B_p} is the proton cyclotron frequency, ω_{12} the crossover frequency at which both electron and proton whistlers propagate, and f_c is the highest frequency for which polarization reversal occurs. Taken from Gurnett, *et al.*,[52] 1965.

propagate. At this frequency polarization reversal of a wave travelling in a slowly varying medium changes from an up-going right-hand-polarized electron whistler to a left-hand polarized proton whistler. When strong coupling exists between the two modes of propagation both waves can propagate to higher altitudes. Sometimes the electron whistler is strongly attenuated and appears cut-off at the frequency f_c. The proton whistler frequency increases with time and approaches the proton cyclotron frequency f_{B_p} at long times.

The He$^+$ and O$^+$ ions attenuate the radio signals at frequencies below their cyclotron frequencies. At low altitudes where the He$^+$ ion densities are high the radio signals at frequencies less than the He$^+$ cyclotron frequency are absorbed. The wave can only reach the high altitudes via the electron whistler mode. Then under polarization

E

reversal at the cross over frequency ω_{12} it becomes a proton whistler. Since ω_{12} approaches f_{B_p} at an altitude of 440 km at local midnight and at 640 km at local noon, proton whistlers are absent below 440 km.

The ratio of the abundance of H^+ ions to electrons can be obtained from the ratio ω_{12}/f_{B_p}. The experimental ratio H^+/e is 0 at 440 km and increases rapidly to about one at 1,400 km at night and more slowly to one at 2,400 km during the day.[52]

The ion temperature is obtained from the cyclotron damping of the proton whistlers. The resulting ion temperature[53] is proportional to the third power of the difference Δf_c between the wave frequency at cut-off and the proton cyclotron frequency. A typical value of Δf is 3 c/s. The proton cut off frequency is shown in Fig. 3.23b. At mid-latitudes at night the temperatures were found to vary from 600° K to 1,050° K during winter and to be about 750° K during summer. These temperatures are in reasonable agreement with the neutral gas temperatures discussed in Chapter 2 on the earth's atmosphere and with the ion scale heights determined from plasma probes at lower altitudes.

Considerable ionospheric information has been extracted from the very low frequency radio detection experiments on satellites. From the proton whistler alone, the electron density, the proton abundance, the magnetic field strength and the proton temperature were obtained.

Ion temperatures were also measured[54] with a retarding potential analyzer on Imp 2 (Explorer 21) that was launched on October 4, 1964 to an apogee of 15 earth radii. The electrons had a Maxwellian temperature distribution.

3.6 IONOSPHERIC WINDS

The variation of the earth's magnetic field during the day and night and the disturbances during magnetic storms have for many years been attributed to large currents flowing in the lower ionosphere. The solar diurnal variations S_D during magnetically quiet times is found by subtracting the average magnetic field on the earth from the instantaneous field. The storm time variation D_{st} is the varying field obtained during magnetic storms. In a similar way the solar diurnal and storm time variations are obtained for the characteristic frequency and height of the maximum of the F region.

One explanation for the driving force for the ionospheric currents was suggested by Martyn.[55] He proposed that all ionospheric disturbance variations are due to electrostatic fields that develop in the auroral zones and spread over the earth throughout the ionosphere. The ion drifts in the F region are caused by electric fields that are communicated upward from the E region. The electric fields drive the ionospheric currents and produce vertical and horizontal diurnal drifts of the ionization. In crossed electric and magnetic fields the drift velocity is given by

$$\mathbf{v} = \frac{c(\mathscr{E} \times \mathbf{B})}{B^2} \cdot \tag{3.33}$$

At a latitude where the eastward electric field is $\mathscr{E}$ and the magnetic field is $\mathbf{B}$ with inclination ψ and where the ion collision frequency is small compared to the cyclotron frequency, the southward drift velocity v_σ and the vertical upward drift velocity v_{up} of the ions and electrons in the $F2$ region is given by

$$v_s = \frac{-c\mathscr{E}\sin\psi}{B}$$

$$v_{\mathrm{up}} = \frac{c\mathscr{E}\cos\psi}{B} \cdot \tag{3.34}$$

The electric field produces a current density j (amps/cm²) almost parallel to $\mathscr{E}$

$$\mathbf{j} = \sigma\mathscr{E}. \tag{3.35}$$

A force/cm³ is exerted by the magnetic field on the current density

$$\mathbf{F} = \frac{\mathbf{j} \times \mathbf{B}}{c} \cdot \tag{3.36}$$

Substituting $\rho\mathbf{a}$ for $\mathbf{F}$ where $\mathbf{a}$ is the acceleration and ρ is the atmospheric density and $\sigma\mathscr{E}$ for $\mathbf{j}$, the horizontal north acceleration becomes

$$a = \frac{\sigma B^2 v_\sigma}{\rho c^2} \cdot \tag{3.37}$$

If the acceleration is impressed for $c^2\rho/\sigma B^2$ sec, about three hours, the entire mass of the atmosphere of the $F2$ region is set in motion. The effect of the motion is to generate a "dynamo" field that is opposed to $\mathscr{E}$ and therefore reduces the drift velocity of the particles. In general the acceleration increases with height as ρ decreases. However, at great heights molecular viscosity and heat conductivity damp the air motion so that the acceleration eventually approaches zero.

Ionospheric electric fields perpendicular to the magnetic field lines of 1 to 3 millivolt/m have recently been measured.[56] Artificial Ba ion clouds were released at altitudes of 100 to 1,000 km and the velocities of the ion clouds were measured by time spaced photographs. In the late morning and early evening the directions of the electric vectors were found to be opposite to the S_D directions. These measurements support the dynamo theory.[55]

On the other hand, it has been suggested[57] that the atmospheric pressure gradients cause atmospheric winds which then produce ionospheric drifts. Opposing the pressure gradient is the ion drag. The ions that are tied to the magnetic field lines collide with the neutral atmospheric atoms and retard their motion. A south–north horizontal component of the atmospheric wind of 34 m/sec was found at middle latitudes. The vertical drift was 20 m/sec. It would appear that this mechanism competes favorably with electric fields for driving the atmosphere.

3.7 WAVE PARTICLE INTERACTIONS

Several recent calculations suggest a strong coupling between the electron whistlers and electrons and between ion cyclotron waves and low energy protons trapped in the magnetosphere.[58-61] The conditions on the frequencies and polarizations are given in Table 3.4. The trapped

Table 3.4 Conditions on frequency and polarization
of ion cyclotron and electron whistler waves

Wave	Frequency	Polarization	Resonant particle
Ion cyclotron	$f \ll f_{B_p}$	Left	Proton
Electron whistler	$f_{B_p} \ll f < |f_{B_e}|$	Right	Electron

particles and the waves travel on paths along the earth's magnetic field lines. When the particles run into the waves the waves' frequencies are Doppler shifted to higher frequencies. If the wave frequency is shifted up to the particle cyclotron frequency, resonance occurs and energy is fed from the particles into the waves. The waves are amplified. This wave amplification is the inverse of the hydromagnetic wave loss mechanism, discussed in Chap. 1. There the electron and proton pitch angles were decreased and the particles were forced to mirror deeper into the atmosphere where they were lost. These predicted amplified waves may have been observed experimentally as the pulse trains of 0.1 to 5 c/s at middle and low latitudes.[62]

An additional condition on the resonant cyclotron instability is that the particle pitch angle distribution must be sufficiently anisotropic. When more particle energy is perpendicular than parallel to the magnetic field lines the energy in the waves grows. This is the usual case when there are atmospheric losses. Pitch angle diffusion changes particles from large to small pitch angles and reduces the anisotropy.

The energy lost to the escaping particles and waves must be replenished. Other particles must be accelerated by some unspecified mechanism to higher energies to furnish the lost particle and wave energies. If the wave growth rate becomes too large, the wave energy dominates, the particle pitch angles are rapidly decreased, and the particles are quickly lost. By this means the growth rate is controlled and a steady state condition is reached.

The amplitude of a wave at time t that moves in the z direction with an angular frequency ω and wave number k with initial amplitude a_0 is

$$a = a_0 \, e^{i(kz-\omega t)}. \tag{3.38}$$

The wave can grow rapidly if ω is positive imaginary. So the standard procedure to find out if the wave will grow is to calculate the index of refraction $\mu = ck/\omega$ and locate a positive imaginary part to ω (for real k).

On doing this Cornwall[58] found that the frequencies for the ion cyclotron waves should be about

$$f = f_{Bp} v_A / v. \tag{3.39}$$

Substituting for the Alfvèn velocity $v_A = B/\sqrt{(4\pi N M_p)}$, where $f_{Bp} = eB/M_p c$ and M_p is the mass of the proton, and taking $v = 10^9$ cm/sec for protons of 500 keV, $B = 0.3/L^3$ gauss and $N = 6 \times 10^3/L^3$ ions/cm^3, we find

$$f = 3 \times 10^2/L^{9/2} \text{ c/s}. \tag{3.40}$$

At $L = 3.5$ this gives $f = 1$ c/s.

The same considerations are true of the electron whistlers with the exception that the electron mass replaces the proton mass in v_A and f_{B_p}. Taking the electron velocity as c and evaluating again at $L = 3.5$, we find $f = 3$ kc/s which falls in the VLF radio band.

Comparisons[59] of the experimental maximum fluxes of electrons with energies greater than 40 keV and protons with energies greater than 120 keV with the calculations indicate that the theoretical maximum flux is seldom exceeded. For $L > 4$ the fluxes are close to the calculated limit. The observed distribution of particles within and near the loss cone is consistent with the wave-particle interactions. The typical amplitude of magnetic field noise produced at the equator near $L = 6$ is about $10^{-2}\gamma$.

This cannot be the whole story because trapped electron fluxes sometimes decrease at values well below the self-excitation limit.[63] And the very interesting time-dependent phenomena and structure in the frequency distribution of the waves has not been discussed.

3.8 SUMMARY OF IONOSPHERE ELECTRON DENSITIES

It is useful to combine the data obtained by the different techniques and to summarize the ionospheric electron density dependence on altitudes. It must be kept in mind that the electron densities vary during the day and night, with the solar cycle, and with solar magnetic storms. Gringauz[64] has given a good summary of the Russian work to 1962, and Bordeau[14] an excellent survey to 1965.

The electron density is summarized in Fig. 24 out to 10 earth radii on a log-log plot. Daytime data from the dispersive Doppler effect,[20,21,65] Faraday rotation,[64,66] plasma probes,[32] ion traps,[50,64] and whistlers[48,49] have been combined to give the curves of Fig. 3.24. The daytime electron density of about 10^2 electrons/cm³ becomes measureable at 80 km in the D region and rises rapidly through the D and E regions to a maximum in the F region of 10^6 electrons/cm³ at 300 km. Beyond the maximum the density gradually falls to 10^3 electrons/cm³ at the knee at three earth radii where it suddenly drops off more rapidly. Beyond the knee it continues to decrease gradually to 1 electron/cm³ at seven earth radii and continues at this low value to the boundary of the magnetic cavity located at 10 earth radii in the sun direction. The density there is lower than the 4 electrons/cm³ observed in interplanetary space. This minimum in ionization in the outer region of the magnetic cavity is sometimes called the electron density trough.

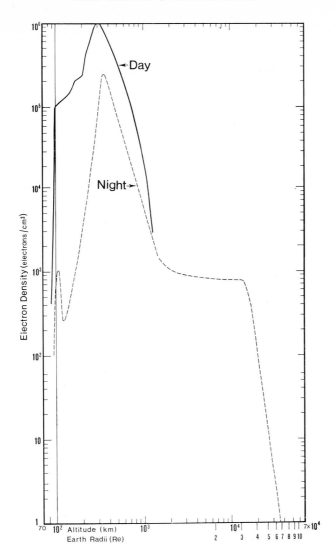

Figure 3.24. Average curves for electron density versus altitude up to 10 earth radii. The density during the day is shown with the solid curve and during the night with the dashed curve. The curves were obtained with data from the dispersive Doppler effect, Faraday rotation, plasma probes, ion traps, and whistlers.

The electron density at night is from one to two orders of magnitude lower than the density during the day at low altitudes. Above the F layer the density at night approaches the density during the day.

Whistler measurements below the knee at three earth radii show little day–night difference. Outside the knee the differences are larger because the knee moves inward about one earth radius during the night.

There is every reason to believe that charge neutrality is preserved and that the positive ion densities follow curves identical to the electrons.

REFERENCES

1. Kennelly, A. E., "On the elevation of the electrically-conducting strata of the earth's atmosphere," *Electrical World and Engineer* **39**, 473 (1902).
2. Heaviside, O., *Encyclopedia Britannica, Tenth Edition* **13**, 215 (1902).
3. Appleton, E. V., "Wireless studies of the ionosphere," *Proc. Inst. Elec. Engr.* **7**, 642–650 (1932).
4. Lorentz, H. A., "*Theory of Electrons*," p. 305, G. E. Stehert and Co., Publishers, New York, New York, 1909.
5. Eccles, W. H., "On the diurnal variations of the electric waves occurring in nature, and on the propagation of electric waves round the bend of the earth," *Proc. Roy. Soc. London* **87A**, 79–99 (1912).
6. Larmor, J., "Why wireless electric rays can bend round the earth," *Phil. Mag. and J. Sci.* **48** (6), 1025–1036 (1924).
7. Breit, G. and Tuve, M. A., "A radio method of estimating the height of the conducting layer," *Nature* **116**, 357 (1925).
8. Kirby, S. S., Berkner, L. V. and Stuart, D. M., "Studies of the ionosphere and their application to radio transmission," *Proc. Inst. Radio Engr.* **22**, 481–521 (1934).
9. Appleton, E. V., "Two anomalies in the ionosphere," *Nature* **157**, 691 (1946).
10. Barkhausen, H., "Zwei mit Hilfe der neuen Verstärker entdeckte Erscheinungen," *Physik, Zeitschr.* **20**, 401–403 (1919).
11. Eckersly, T. L., "A note on musical atmospheric disturbances," *Phil. Mag.* **49**, 1250–1260 (1925).
12. Burton, E. T. and Boardman, E. M., "Audio-frequency atmospherics," *Proc. Inst. Radio Engrs.* **21**, 1476–1494 (1933).
13. Storey, L. R. O., "An investigation of whistling atmospherics," *Phil. Trans. Roy. Soc.* **246**, 113–141 (1953).
14. Bourdeau, R. E., "Research within the ionosphere," *Science* **148**, 585–594 (1965).
15. Nicolet, M. and Aikin, A. C., "The formation of the D region of the ionosphere," *J. Geophys. Res.* **65**, 1469–1483 (1960).
16. Kreplin, R. W., Chubb, T. A. and Friedman, H., "X-ray and Lyman-Alpha emission from the sun as measured from the NRL SR-1 satellite," *J. Geophys. Res.* **67**, 2231–2253 (1962).
17. Maehlum, B. and O'Brien, B. J., "Solar cosmic rays of July 1961 and their ionospheric effects," *J. Geophys. Res.* **67**, 3269–3279 (1962).
18. Watanabe, K. and Hinteregger, H. E., "Photoionization rates in the E and F regions," *J. Geophys. Res.* **67**, 999–1006 (1962).

19. Taylor, H. A., Jr. and Brinton, H. C., "Atmospheric ion composition measured above Wallops Island, Virginia," *J. Geophys. Res.* **66**, 2587–2588 (1961).
20. Seddon, J. C. and Jackson, J. E., "Ionosphere electron densities and differential absorption," *Annales de Geophysique* **14**, 456–463 (1958).
21. Gringauz, K. I., "Rocket measurements of the electron concentration in the ionosphere by means of an ultrashort-wave dispersion interferometer," *Soviet Physics Dokl.* **3**, (3), 620–623 (1958).
22. Allen, C. W., "World-wide diurnal variations in the $F2$ region," *J. Atmos. Terrest. Physics* **4**, 53–67 (1955).
23. Munro, G. H., "Travelling disturbances in the ionosphere: diurnal variation of direction," *Nature* **171**, 693 (1953); "Travelling disturbances in the ionosphere," *Proc. Roy. Soc. London* **202A**, 208–223 (1954).
24. Appleton, E. V., "The anomalous equatorial belt in the $F2$ layer," *J. Atmos. Terrest. Physics* **5**, 348–351 (1954).
25. Croom, S., Robbins, A. and Thomas, J. O., "Two anomalies in the behavior of the $F2$ layer of the ionosphere," *Nature* **184**, 2003–2004 (1959).
26. Rastogi, R. G., "The diurnal development of the anomalous equatorial belt in the $F2$ region of the ionosphere," *J. Geophys. Res.* **64**, 727–731 (1959).
27. Sayers, J., Rothwell, P. and Wager, J. H., "Field aligned strata in the ionization above the ionospheric $F2$ layer," *Nature* **198**, 230–233 (1963).
28. King, J. W., "Investigations of the upper ionosphere deduced from top-side sounder data," *Nature* **197**, 639–641 (1963).
29. King, J. W., Smith, P. A., Eccles, D., Fooks, G. F. and Helm, H., "Preliminary investigation of the structure of the upper ionosphere as observed by the topside sounder satellite, Alouette," *Proc. Roy. Soc. London* **281A**, 464–487 (1964).
30. King, J. W., Eccles, D., Legg, A. J., Smith, P. A., Galindo, P. A., Kaiser, B. A., Preece, D. M. and Rice, K. C., "An explanation of various ionospheric and atmospheric phenomena including the anomalous behavior of the F region," *Radio Research Station, Ditton Park, Slough, England, Document No. RRS/I.M. 191* (1964).
31. Thomas, J. O. and Sader, A. Y., "Electron density at the Alouette orbit," *J. Geophys. Res.* **69**, 4561–4581 (1964).
32. Brace, L. H. and Spencer, N. W., "First electrostatic probe results from Explorer 17" *J. Geophys. Res.* **69**, 4686–4689 (1964).
33. Willmore, A. P., "Geographical and solar activity variations in the electron temperature of the upper F region," *Proc. Roy. Soc. London* **286**, 537–558 (1965).
34. Dalgarno, A., McElroy, M. B. and Moffett, R. J., "Electron temperatures in the ionosphere," *Planet. Space Sci.* **11**, 463–484 (1963).
35. Brace, L. H., Reddy, B. M. and Mayr, H. G., "Global behavior of the ionosphere at 1,000 kilometer altitude," *J. Geophys. Res.* **72**, 265–283 (1967).
36. Hanson, W. B., "Electron temperatures in the upper atmosphere," *Space Res. III*, pp. 282–302, (ed.), Priester, North-Holland Publishing Co., Amsterdam, 1963.
37. Hanson, W. B. and McKibbin, D. D., "An ion-trap measurement of the ion concentration profile above the $F2$ peak," *J. Geophys. Res.* **66**, 1667–1671 (1961).
38. Bourdeau, R. E., Whipple, E. C., Jr., Donley, J. L. and Bauer, S. J., "Experimental evidence for the presence of helium ions based on Explorer VIII satellite data," *J. Geophys. Res.* **67**, 467–475 (1962).

39. Bowen, P. J., Boyd, R. L. F., Raitt, W. J. and Willmore, A. P., "Ion composition of the upper F region," *Proc. Roy. Soc. London* **281**, 504–514 (1964).

40. Barrington, R. E., Belrose, J. S. and Nelms, G. L., "Ion composition and temperatures at 1,000 km as deduced from simultaneous observations of a VLF plasma resonance and topside sounding data from the Alouette 1 satellite," *J. Geophys. Res.* **70**, 1647–1664 (1965).

41. Nelms, G. L., "Ionospheric results from the topside sounder satellite Alouette," *Space Research IV*, pp. 437–448, North-Holland Publishing Co., Amsterdam, 1963.

42. Gledhill, J. A., Torr, D. G. and Torr, M. R., "Ionospheric disturbance and electron precipitation from the outer radiation belt," *J. Geophys. Res.* **72**, 209–214 (1967); Gledhill, J. A. and Torr, D. G., "Ionospheric effects of precipitated electrons in the south radiation anomaly," *Space Research VI*, pp. 222–229, North-Holland Publishing Co., Amsterdam, 1965.

43. Bates, H. F., Belon, A. E., Romick, G. J. and Stringer, W. J., "On the correlation of optical and radio auroras," *J. Atmos. Terrest. Physics* **28**, 439–446 (1966).

44. Helliwell, R. A., Crary, J. H., Pope, J. H. and Smith, R. L., "The 'nose' whistler—a new high-latitude phenomenon," *J. Geophys. Res.* **61**, 139–142 (1956).

45. Smith, R. L., "Properties of the outer ionosphere deduced from nose whistlers," *J. Geophys. Res.* **66**, 3709–3716 (1961).

46. Carpenter, D. L., "Whistler evidence of a 'knee' in the magnetospheric ionization density profile," *J. Geophys. Res.* **68**, 1675–1682 (1963).

47. Carpenter, D. L. and Smith, R. L., "Whistler measurements of electron density in the magnetosphere," *Rev. Geophys.* **2**, 415–441 (1964).

48. Carpenter, D. L., "Whistler studies of the plasmapause in the magnetosphere—1. Temporal variations in the position of the knee and some evidence on plasma motions near the knee," *J. Geophys. Res.* **71**, 693–709 (1966).

49. Angerami, J. J. and Carpenter, D. L., "Whistler studies of the plasmapause in the magnetosphere—2. Electron density and total tube electron content near the knee in magnetospheric ionization," *J. Geophys. Res.* **71**, 711–725 (1966).

50. Gringauz, K. I., Kurt, V. G., Moroz, V. I. and Shklovskii, I. S., "Result of observations of charged particles observed out to $R = 100,000$ km with the aid of charged particle traps on soviet space rockets," *Soviet Astron. AJ* **4**, 680–695 (1960).

51. Taylor, H. A., Jr., Brinton, H. C. and Smith, C. R., "Positive ion composition in the magnetoionosphere obtained from the Ogo-A satellite," *J. Geophys. Res.* **70**, 5769–5781 (1965).

52. Gurnett, D. A., Shawhan, S. D., Brice, N. M. and Smith, R. L., "Ion cyclotron whistlers," *J. Geophys. Res.* **70**, 1665–1688 (1965).

53. Gurnett, D. A. and Brice, N. M., "Ion temperature in the ionosphere obtained from cyclotron damping of proton whistlers," *J. Geophys. Res.* **71**, 3639–3652 (1966).

54. Serbu, G. P. and Maier, E. J. R., "Low energy electrons measured on Imp 2," *J. Geophys. Res.* **71**, 3755–3766 (1966).

55. Martyn, D. F., "The morphology of the ionospheric variations associated with magnetic disturbance—1. Variations at moderately low latitudes," *Proc. Roy. Soc. London* **218**, 1–18 (1953).

56. Haerendel, G., Lüst, R. and Reiger, E., "Ionospheric electric fields at mid-latitudes during twilight," unpublished report, Max-Planck-Institüt für Physik und Astrophysik, Institüt für Extraterrestische Physik, Munich, Germany.

57. King, J. W. and Kohl, H., "Upper atmospheric winds and ionospheric drifts caused by neutral air pressure gradients," *Nature* **206**, 699–701 (1965).

58. Cornwall, J. M., "Cyclotron instabilities and electromagnetic emission in the ultra low frequency and very low frequency ranges," *J. Geophys. Res.* **70**, 61–69 (1965).

59. Obayashi, T., "Hydromagnetic whistlers," *J. Geophys. Res.* **70**, 1069–1078 (1965).

60. Kennel, C. F. and Petschek, H. E., "Limit on stably trapped particle fluxes," *J. Geophys. Res.* **71**, 1–28 (1966).

61. Liemohn, H. B., "Cyclotron-resonance amplification of VLF and ULH whistlers," *J. Geophys. Res.* **72**, 39–55 (1967).

62. Tepley, L., "Regular oscillations near 1 c/s observed at middle and low latitudes," *Radio Sci. J. Res.* NBS/USNC-URSI **69D** (8), 1089–1105 (1965).

63. Frank, L. A., Van Allen, J. A. and Hills, H. K., "An experimental study of charged particles in the outer radiation zone," *J. Geophys. Res.* **69**, 2171–2191 (1964).

64. Gringauz, K. I., "The structure of the ionized gas envelope of earth from direct measurements in the U.S.S.R. of local charged particle concentrations." *Planet. Space Sci.* **11**, 281–296 (1963).

65. Jackson, J. E. and Bauer, S. J., "Rocket measurement of a daytime electron-density profile up to 620 kilometers," *J. Geophys. Res.* **66**, 3055–3057 (1961).

66. Bauer, S. J. and Jackson, J. E., "Rocket measurement of the electron density distribution in the topside ionosphere," *J. Geophys. Res.* **67**, 1675–1676 (1962).

CHAPTER 4

The Sun

4.1 INTRODUCTION

The sun is only one of some 10^{20} stars in our observable universe. But it is the most important star because it is closer to us than any other star by a factor of 300,000. The ancient civilizations all recognized their dependence on the sun. The Egyptians 3,000 years ago believed that the sky goddess Mut swallowed the setting sun. It traveled through her body during the night and each morning was born again at dawn.

In Homer's Odyssey the vengeance of the sun was swift. Odysseus and his men were resting on the island of the sun after a vigorous voyage. While he was away praying, his crew became hungry and killed the sacred oxen. When the men left the island the sun sent a thunderbolt that shattered the ship and all were drowned but Odysseus.

The earth was thought to be the center of the universe until Copernicus published his De revolutionibus orbium coelestius in 1592. He made the bold suggestion that the earth and planets revolve around the sun, the center of the solar system.

Our sun is much like many other stars. It is 3×10^{16} km from the center of our spiral galaxy and rotates at 200 km/sec about that center. It has a radius of 6.96×10^5 km and its diameter subtends an angle of 1/2 deg at the earth. It has a mass of 2.0×10^{33} gm and a mean density of 1.5 gm/cm³. Its core density is very high, 110 gm/cm³, while its density at one optical depth (one mean-free path for light absorption) is only 10^{-7} gm/cm³.

The sun's gaseous atmosphere rotates faster at the equator than at the poles. The sunpsot rotation periods, measured in white light from the photosphere, change from 25 days at the equator to 30 days at the poles. This variation of the period is probably responsible for the

winding up of the sun's magnetic fields and the 22 year cycle in the
sun's magnetic field polarity.

A sketch of the sun is shown in Fig. 4.1. The cut-away shows the
interior. Because of light and radio absorption no observations have
been made below the photosphere, the layer that is seen with a telescope
in white light. The internal structure of the sun is deduced purely
from theoretical considerations. Hydrogen nuclear reactions take
place in the center of the sun where the temperature is 10^7 °K and the
pressure is 10^{11} atmospheres. The energy released when 4 hydrogen
nuclei are combined to form one He nucleus heats the sun. The
reactions take place within 0.2 sun radii $R_\odot$ of the center of the sun.
The energy is transported toward the surface by radiation. The free
and bound electrons absorb and emit the X-rays many times in the
radiation region as the energy works its way out to $0.8R_\odot$. From there
to the surface the energy is transported by convection. Bubbles of gas
are formed that are buoyant and move upward. The convection

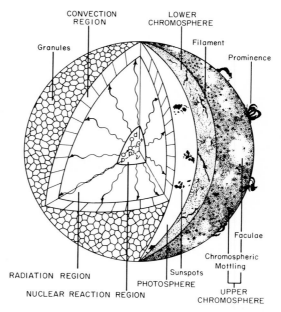

Figure 4.1. Sketch of the sun. A section is cut away to show the nuclear reactions
in the interior, the radiation zone and the convection zone. The granules are
sketched on one section of the photosphere, sunspots on another. Filaments are
shown in the lower chromosphere and faculae, prominences and chromospheric
mottling in the upper chromosphere.

Table 4.1 A model solar atmosphere constructed by Unsöld[1]

Height z, km	Solar radii	Temperature T, °K	Gas pressure, dynes/cm², $\log P$	Electron pressure, dynes/cm², $\log P_e$	Electrons/cm³ $\log N_e$	Turbulent velocity V_t or ΔV, km/sec	Layer	Main energy transfer
1,400,000	3.0	$2\cdot10^6$	-3.8	-4.1	5.5			
700,000	2.0	$2\cdot10^6$	-2.8	-3.1	6.4		Corona	Thermal conduction
350,000	1.50	$2\cdot10^6$	-2.1	-2.4	7.2			
42,000	1.06	$2\cdot10^6$	-0.9	-1.2	8.4			
20,000	1.03	$2\cdot10^6$	-0.8	-1.1	8.5			Mechanical energy
		← Very inhomogeneous →				~15	Transition layer	
3,000	***	~4–6000	0.2	-1.7	10.5			Radiation
2,000		~4–6000	0.5	-1.4	10.8	12	Chromosphere	
1,000		< 4–6000	1.2	-0.9	11.3	7		
	Opt. depth τ_{5000}							
0	0.005	4090	4.1	-0.5	11.7	1–2		
	0.01	4295	4.3	-0.3	12.0			
	0.05	4855	4.6	$+0.2$	12.4		Photosphere	Radiation
	0.1	5030	4.8	$+0.4$	12.6			
	0.5	5805	5.1	1.2	13.3	2		
	1.0	6400	5.2	1.8	13.8			
-260	2.0	7180	5.3	2.4	14.4		Hydrogen Convection zone	Convection
-280	***	10^4	5.3	4.0	15.86	2		
	Solar radii							
$-16,000$	-0.02	10^5	9.4	9.1	20.0	0.3		Radiation
$-140,000$	-0.2	10^6	12.3	12.0	21.9	0.0		

granules seen in the photosphere are sketched on the left half of the sun.

The photosphere is the layer of the sun, 300 km thick, that can be seen with visible light. It contains the sunspots that are illustrated in Fig. 4.1. The right-hand side of the figure shows three strips taken at different altitudes: (1) in the photosphere, (2) 1,000 km above in the lower chromosphere as seen in hydrogen-alpha light, and (3) a few thousand km above the photosphere in the upper chromosphere as seen in the calcium K line. Note the filaments in the lower chromosphere and the faculae and chromospheric mottling in the upper chromosphere. Prominences on the limb can sometimes be seen that splash out as far as 100,000 km. The chromosphere merges into the corona at about 10,000 km above the photosphere.

The solar atmosphere of Unsöld[1] is given in Table 4.1. The heights are measured from the top of the photosphere. The temperature decreases from 2×10^7 °K in the core to 10^6 °K at the outer edge of the radiation region. Through the convection zone the temperature drops to 6,000° K at the bottom of the photosphere, reaches a minimum of 4,000° K at the top of the photosphere and then rises rapidly to 2×10^6 °K in less than 20,000 km in the transition region between the chromosphere and the corona. At higher altitudes the temperature stays constant, but the gas and electron density continue to decrease out into the corona. The main energy transfer mechanisms in the various regions are also indicated in Table 4.1.

4.2 ABUNDANCES OF THE ELEMENTS ON THE SUN

The abundances of the elements on the sun have been measured using the intensities of the absorption or Fraunhofer lines. These abundances have been studied by Goldberg, Muller, and Aller[2] and are compared to earlier investigations in Table 4.2. There the $\log(10^{12}N_{el}/N_H)$ is listed. N_{el} is the number density of the element and N_H is the number density of hydrogen. Helium is not listed because no suitable lines were available for analysis; however, its abundance is thought to be about 0.15 that of hydrogen. The C, N, O group is about 10^{-3} of H. Si and Fe are 5×10^{-5} and 5×10^{-6} of H respectively.

The sun and planets were probably condensed from interstellar gaseous material about 5 billion years ago. The sun is about 70 per cent hydrogen by weight with small amounts of almost all of the elements. Helium must be almost as abundant as hydrogen in the core.

Table 4.2 Comparison of solar abundances[2]

Atomic number	Element	This investigation (1960)	Claas (1951)	Unsöld (1948)	Goldberg–Aller (1943)	Russell (1929)
					$\log(N_{el}/N_H)$	
1	H	12.00	12.00	12.00	12.00	11.5
3	Li	0.96	1.08	--	—	2.0
4	Be	2.36	—	—	—	1.8
6	C	8.72	—	8.29	7.56	7.4
7	N	7.98	—	8.61	8.09	7.6
8	O	8.96	8.65	8.73	8.56	9.0
11	Na	6.30	6.33	6.28	6.56	7.2
12	Mg	7.40	7.57	7.51	8.39	7.8
13	Al	6.20	6.17	6.33	6.39	6.4
14	Si	7.50	7.12	7.29	7.87	7.3
15	P	5.34	—	—	—	—
16	S	7.30	—	6.92	7.57	5.7
19	K	4.70	5.01	5.20	5.09	6.8
20	Ca	6.15	6.46	6.23	6.57	6.7
21	Sc	2.82	—	3.33	—	3.6
22	Ti	4.68	7.56	4.96	4.57	5.2
23	V	3.70	—	4.05	4.09	5.0
24	Cr	5.36	—	5.58	4.87	5.7
25	Mn	4.90	—	5.46	5.09	5.9
26	Fe	6.57	7.16	7.26	6.99	7.2
27	Co	4.64	—	5.03	4.69	5.6
28	Ni	5.91	—	5.95	6.39	6.0
29	Cu	5.04	4.80	4.23	4.39	5.0
30	Zn	4.40	4.52	4.78	5.57	4.9
31	Ga	2.36	—	—	—	2.0
32	Ge	3.29	—	—	—	3.0
37	Rb	2.48	—	—	—	1.7
38	Sr	2.60	2.88	3.35	—	3.3
39	Y	2.25	—	3.21	—	2.6
40	Zr	2.23	—	2.37	—	2.5
41	Nb	1.95	—	—	—	1.0
42	Mo	1.90	—	1.78	—	1.4
44	Ru	1.43	—	—	—	1.7
45	Rh	0.78	—	—	—	0.5
46	Pd	1.21	—	—	—	1.1
47	Ag	0.14	—	—	—	1.0
48	Cd	1.46	—	—	—	2.2
49	In	1.16	—	—	—	0.0
50	Sn	1.54	—	—	—	1.2
51	Sb	1.94	—	—	—	0.8
56	Ba	2.10	2.38	2.95	—	3.3
70	Yb	1.53	—	—	—	—
82	Pb	1.33	—	2.55	—	1.2

The two Burbidges, William Fowler and Fred Hoyle,[3] described the build-up of the elements in the stars in a classic paper in 1957. The elements were formed by proton, helium and neutron reactions. At equilibrium the core temperature adjusts so that the outflow of energy from the star's surface is just balanced by the internal nuclear energy generated. The internal gas pressure is just balanced by the contraction pressure of the force of gravity.

In the contraction, the gravitational energy is converted half into internal thermal energy and half into energy that escapes from the surface.[4] At equilibrium

$$2U + V_g = 0 \qquad (4.1)$$

where U is the internal thermal energy and V_g is the gravitational energy. In the case of the sun the contraction stopped when the temperature reached about 2×10^7 °K. Then the internal kinetic energies of the hydrogen nuclei were sufficiently high for pp reactions to take place and furnish the nuclear energy. Two protons react to form a deuteron, a positron and a neutrino by the reaction

$$p + p \rightarrow d + \beta^+ + \nu \qquad (4.2)$$

The cross section (probability for the reaction to take place) is so small that the reaction has never been measured in the laboratory. Nevertheless astrophysicists have great confidence in its validity. The resultant deuteron then captures a proton to form He^3 and a gamma ray.

$$p + d \rightarrow He^3 + \gamma \qquad (4.3)$$

Finally two of the resultant He^3 nuclei react to form He^4 and 2 protons.

$$He^3 + He^3 \rightarrow He^4 + 2p \qquad (4.4)$$

In the process 4 protons were combined into one He^4 nucleus plus two positrons. The net release of energy is 26 MeV.

The total energy release from the sun per sec can now be estimated. The sun's energy flux measured at the surface of the earth is 1.95 cal/cm²-min (1.36×10^6 erg/cm²-sec). The mean distance from the earth to the sun is 1.50×10^8 km so that the total energy emitted by the sun is found to be 3.9×10^{33} erg/sec. Every second 7.5×10^8 ton of hydrogen converts to He^4. This energy is equivalent to 10^{11} hydrogen bombs of 1 megaton yield each exploding every second. Still the sun can continue at the present rate of hydrogen fuel consumption for another 5 billion years before converting to helium burning.

The capture of protons by the heavier elements with higher nuclear charge and the burning of helium and higher nuclear charge elements requires higher temperatures. Since He has a charge $Z = 2$, C has $Z = 6$ and Fe has $Z = 26$, helium burning requires a higher temperature than hydrogen, carbon higher than helium and iron higher than carbon. The energy necessary to overcome the repulsion of the nuclear charge is the Coulomb energy

$$V_c = \frac{Z_1 Z_2 e^2}{r} \tag{4.5}$$

where Z_1 and Z_2 are the charges of the two reacting nuclei, e is the electron or proton charge of 4.8×10^{-10} e.s.u. and r is the separation distance of the two nuclei. It is possible for the particles to "tunnel through" the Coulomb barrier, although with a lower probability, even if the kinetic energies of the reacting particles are not sufficiently high to surmount the Coulomb barrier.

The automatic temperature rise accompanied by contraction and conversion of gravitational energy to thermal energy stops at Fe. For elements with higher nuclear charge no net energy release is obtained from combination or fusion of the light elements. The higher Z elements can be built from neutron capture with subsequent β decay. Slow and rapid n capture occurs. The rapid capture probably takes place in supernovae explosions where transuranic elements as high as Cf^{254} are formed.

4.3 ORIGIN OF THE SUN

There are two opposite theories for the origin of the primordial matter, the single event and the continuous creation. The red shift of the light from the distant galaxies has been interpreted as a Doppler shift caused by the velocities of recession. The velocities may have originated from a primordial "big bang" explosion 10 billion years ago when all the matter of the universe was ejected from a common point. What happened before the great explosion must forever remain a mystery. On the other hand a steady state creation with matter continuously appearing and disappearing throughout the universe is also possible. The steady state universe requires a stellar origin of the elements. The explosive universe, also works well with the stellar element formation, provided the primordial explosion produced sufficient hydrogen.

A star spends most of its lifetime on the main sequence of the Hertzsprung–Russell or luminosity–temperature diagram shown in Fig. 4.2. The sun is a typical young star on that diagram. After using up all the central hydrogen about 5 billion years from now, it is thought that the sun will expand to burn its surface hydrogen. It will convert to He^4 fuel as a red giant. Its core will contract until it reaches a

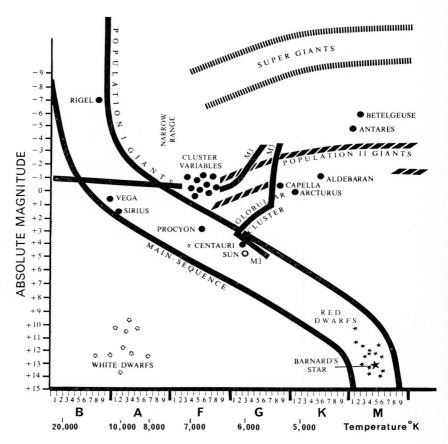

Figure 4.2. A Herzsprung–Russell (H–R) diagram of absolute stellar magnitudes versus spectral color. On the horizontal axis the temperature increases toward the left. The main sequence is between the two heavy lines running from upper left to lower right. Several stellar objects are shown. From L. Motz and A. Duveen, *Essentials of Astronomy*, Wadsworth Publishing Company, Inc., Belmont, California, 1967.

temperature of about 10^8 K where the He^4 nuclei can combine to form C^{12} by the reaction

$$3He^4 \rightarrow C^{12} + \gamma \qquad (4.6)$$

At that time it will likely consume the earth and other planets.

4.4 TRANSPORT OF ENERGY ON THE SUN

No observations have been made of the sun's interior. Visible light originates in the top one gm/cm^2 of the photosphere. Other radiations, ultraviolet, radio and microwave originate even higher in the chromosphere or lower corona so they give no information about the sun's interior. Neutrons, if observed, could come from a few tens of grams below the surface. The only hope of obtaining information about the deep interior, however, is to detect and measure the properties of the elusive neutrinos that penetrate great amounts of material. They can easily escape from the center of the sun. For the same reason they are very difficult to find. Extremely large detectors with very low background counting rates must be used to detect the neutrinos.

At the present time we must depend on theoretical calculations to describe the transport of energy in the sun. Deep inside the sun, energy is transmitted outward by X-ray absorption and emission. Near the surface at $0.8R_\odot$ the transport of energy by convection replaces radiation transport. The turbulent boiling mass motion of the sun becomes important in transporting energy to the surface. There the gas pressure is about 10^{12} dynes/cm^2 and the temperature is about 10^6 °K.[1] The sun is convectively unstable from there up to the lower layer of the photosphere where the pressure is about 10^5 dynes/cm^2 and the temperature is 6,500° K. Below this layer the hydrogen is completely ionized, inside the layer it is partially ionized and above the layer it is mostly neutral.

When a volume of ionized hydrogen gas rises, some of the ions capture electrons to form neutral hydrogen atoms. Each capture releases 13.6 eV of energy. The volume becomes hotter than the surrounding gas, its density less, and buoyant forces cause it to rise. It continues to rise until it reaches the photosphere. By that time it is composed mostly of neutral atoms. In the same way a volume of gas that starts downward continues to move downward until it is totally ionized.

Photographs taken in white light by the Aerospace Corporation San Fernando Observatory 6-inch refracting telescope show very sharp

images of the granules in the photosphere.[5]   See the photograph of
Fig. 4.3. Only under the best "seeing" conditions is it possible to obtain
pictures of the granules from the earth. We can see that the whole
surface of the sun is covered by small white cells surrounded by thin
black lines. Measurements of the similar photographs show a brightness
cell full width at half maximum of 650 km.[6]

The vertical velocities are measured by the Doppler shifts of different
frequencies of light.   They vary from about 0.4 km/sec at the lowest
photosphere altitudes to 1.6 km/sec in the upper photosphere.   The
time for the velocity to rise, fall and then return to its original position
is 5 minutes.[7] The period as a function of altitude was measured using
spectral lines that originate at different heights in the photosphere and
chromosphere.   The period was found to decrease slowly with altitude.
The times of highest intensity correlate with highest vertical velocity.
Hot gases may be moving upward in the center of the cell and cold
gases down on the periphery.

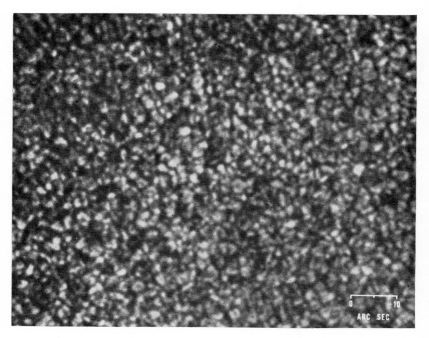

Figure 4.3. Solar granulation photographed in July 1967 with a 6-inch
refracting telescope. From Earle Mayfield, Aerospace Corporation San Fernando
Observatory.

Leighton, Noyes and Simon[7] used a clever technique of overlaying spectroheliograms taken 2.5 min apart to cancel out the bright and dark regions of granules. (A spectroheliograph is a spectrograph that can be moved relative to the fixed sun and the fixed detecting photographic plate in a direction perpendicular to the entrance slit. An image of the sun at the wavelength selected by the spectrometer is pictured on the photographic plate. A drawing of a spectroheliograph is shown in Fig. 4.4.) Supergranulations then appear with a width of about 15,000 km, about 20 times the width of the small granulations. Here the velocities are horizontal instead of vertical. The bright and dark regions appear near the limb of the sun instead of near its center. In each supergranulation, the motion on the side nearest the center is toward the limb. A photograph of the supergranulations is shown in Fig. 4.5. The bright areas are the regions approaching the center of the sun and the dark areas are the regions receding.

The ordered motions of the granulations cause disturbances in the overlying regions. It is thought that a substantial fraction of the energy is propagated into the chromosphere and lower corona as acoustic or

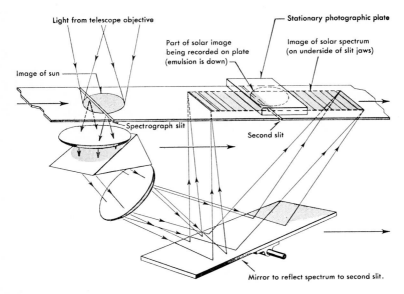

Figure 4.4. Drawing of a spectroheliograph. The entire apparatus moves except for the photographic plate. A spectroheliograph is very useful for photographing the sun at a desired wavelength of light. From Abell, *Exploration of the Universe*, Holt–Rinehart and Winston, New York.

Figure 4.5. Supergranulation on the sun taken from Leighton,[6] 1963. The velocities on the surface of the sun are measured by the Doppler shifts of particular wavelengths of light. The velocity is away from the observer in the dark areas and toward the observer in the light areas. The photograph shows the large supergranulations of 15,000 km width.

pressure waves. These waves are eventually transformed into shock waves that dissipate to heat up the chromosphere and corona.

4.5 THE PHOTOSPHERE

The photosphere of the sun is the layer about 300 km thick that radiates most of the energy received on the earth. The energy is mostly in the visible with its maximum in the green. The photosphere also absorbs radiation and produces the Fraunhofer absorption line spectrum that is superimposed upon the continuous emission spectrum.

The fraction of the radiation that is absorbed in traveling a distance dl (cm) through the sun's atmosphere is

$$dI/I = - k_\nu \rho \, dl \tag{4.7}$$

where ρ is the density of absorbing atoms in gm/cm³ and k_ν is the absorption coefficient in cm²/gm. The frequency of radiation is indicated by ν. The absorption coefficient varies with frequency and has its minimum in the visible as shown in Fig. 4.6. It rises rapidly as the wavelength decreases (frequency or energy increases) and has a maximum at the He alpha wavelength of 280 A (energy of 44 eV). The absorption edges near the resonant frequencies for H, He and Si are labeled.

The light absorption is often measured in terms of optical depths given by

$$\tau_\nu = \int_\infty^r k_\nu \, \rho \, dl \qquad (4.8)$$

where one optical depth is the distance required to reduce the intensity by $1/e$. Consequently it is possible to look down into the sun to only about one optical depth.

The Fraunhofer lines of the neutral metals that are responsible for most of the white light are formed at about 0.1 optical depths down into the photosphere. There the temperature is about 5,000° K and the gas pressure is 9×10^4 dynes/cm².[1] The H Balmer α radiation (6563 A) with a larger absorption coefficient originates in the lower chromosphere. Radiation of the K line of Ca II with a higher absorption coefficient yet, originates in the upper chromosphere.

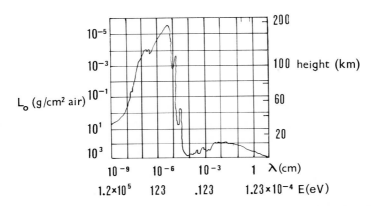

Figure 4.6. A plot of the mean-free path in the atmosphere versus the wavelength of the radiation. From K. S. W. Champion, McGraw–Hill Yearbook of Science and Technology, McGraw–Hill, Inc., N.Y., 1965.

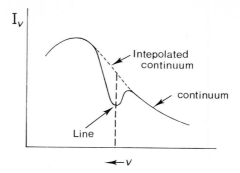

Figure 4.7. Profile of a Fraunhofer line. The continuum, interpolated continuum and line spectrum are shown.

The contour of a Fraunhofer line is shown in Fig. 4.7. The continuum radiation, the interpolated continuum and the decreased intensity in the vicinity of the line are sketched versus frequency.

The absorption is maximum at the natural or resonant frequency of the classical atom. This is the frequency at which the incoming photon has an energy exactly equal to the energy between two levels of the atom. The energy or frequency width of the absorption line is not zero but is broadened for several reasons. The quantum mechanical line broadening is caused by the finite lifetime of the state through the Heisenberg uncertainty principle. If the lifetime is short the frequency

$$\Delta\nu\,\Delta t = 1/2\pi \tag{4.9}$$

or line width is broad. The lines are also broadened by collisions between atoms of the gas. The collisions change the energy states of the emitting atoms slightly. In self-broadening the disturbing atoms are the same as the emitting atoms, in Van der Waal's broadening the atoms are neutral hydrogen and in quadratic Stark broadening the perturbing particles are ions and electrons.

The lines can be shifted by the electrostatic fields of ions and electrons in the linear Stark effect. This results in a statistical broadening of the line as different atoms receive different shifts. The Doppler broadening is caused by the motions of the emitting atoms. Atoms with different velocities and different directions cause a statistical broadening of the lines. Shifts of the lines by directed velocities were used to determine the velocities of the granules in section 4.

4.6 THE CHROMOSPHERE

The chromosphere is the region from 10,000 to 15,000 km thick above the photosphere. At its bottom the temperature is about 4,000° K and the gas pressure is about 10^4 dynes/cm² (see Table 4.1). The first 3,000 km of the chromosphere is composed mostly of cool neutral hydrogen at temperatures from 4,000 to 6,000° K. Above this region the temperature rapidly increases to 2×10^6 °K at 20,000 km. The gas is so rare that the radiation is not absorbed and instead of the Fraunhofer lines, bright emission lines are observed. The strongest is Ca II K observed as high as 10,000 km. The Doppler shifts of the frequencies of the lines show that the chromosphere has a turbulent motion that increases from 12 km/sec at the bottom to 15 km/sec at the top.

The heat energy released in the chromosphere by the turbulent velocity v_t is approximately $v_t{}^3/h$ (ergs/gm-sec) where h is the scale height for the heating.[1] Just above the photosphere where the gas pressure is less than about 10^3 dynes/cm² the heat energy released becomes equal to the energy radiated $k\sigma_S 4T^4$ erg/cm²-sec. k is the average absorption coefficient, σ_S is the Stefan–Boltzmann constant, 5.67×10^{-5} erg/cm²-sec-deg⁴, and T is the temperature. At higher altitudes the heat energy dissipated by the turbulent motion increases rapidly. The temperature then increases to radiate the energy away. Because of the low gas density most of the radiated energy from the photosphere passes through the chromosphere unimpeded. The energy radiated from the chromosphere and corona is small compared to that from the photosphere so is lost in the photosphere background unless the photosphere is blocked out.

For a long time the properties of the chromosphere and the corona, the region beyond the chromosphere, were only obtained at times of solar eclipse when the photosphere was obscured by the moon. After the French astronomer Bernard Lyot invented the coronograph 25 years ago, astronomers no longer needed to wait for an eclipse. An opaque circular diaphragm is placed in the focal plane of the image of the telescope which blocks out the image of the photosphere. A photograph taken with a coronagraph is shown in Fig. 4.8. Tiny hairlike jets called spicules protrude 10,000 km above the limb of the sun. These are probably luminous shock waves propagating through the chromosphere into the corona at supersonic velocities as high as 30 km/sec.

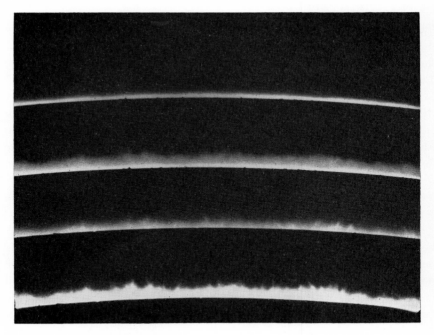

Figure 4.8. Photograph showing spicules in the chromosphere taken with a coronagraph. From Sacramento Peak Observatory, Air Force Cambridge Research Laboratory.

Photographs of the lower and upper chromosphere taken in the light of Hα and Ca K are compared with a white light photograph of the photosphere in Fig. 4.9. The sunspots of the photosphere, the filaments of the lower chromosphere, and the faculae and chromospheric mottling of the upper chromosphere are obvious features of these three regions.

In the transition region in the upper chromosphere the turbulent velocities exceed the velocity of sound. Shock waves or magneto-hydromagnetic waves carry energy outward. Eventually the temperature is so high and the gas density so low that the energy is most efficiently transported outward by thermal conduction of electrons. This region is called the corona.

4.7 THE CORONA

The corona was observed at times of total eclipse at least as far back as Plutarch in A.D. 100. Photographs of the corona were first taken

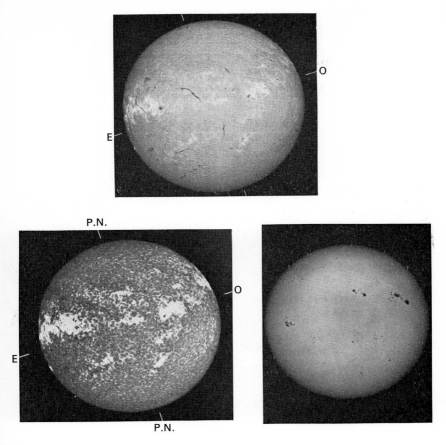

Figure 4.9. (*A*) Picture of the sun in hydrogen-Balmer-α-radiation from the lower chromosphere. (*B*) Picture of the sun taken in the light of Ca *K* from the upper chromosphere. (*C*) Photograph of the sun taken in white light at a different time. The top and bottom left photographs from Observatoire de Paris, Meudon. Bottom right picture from Earle Mayfield, Solar Observatory, Aerospace Corporation, El Segundo, California, 1967.

in the 19th century. These show that the corona extends at least 2 million km from the sun with a light intensity about half that of the moon. The corona is distinguished by the streamers that can be seen out to several sun radii. Photographs of these taken at the maximum, middle and minimum of the solar cycle are shown in Fig. 4.10. At the maximum the streamers are symmetric around the sun and extend outward great distances. At intermediate and minimum times the

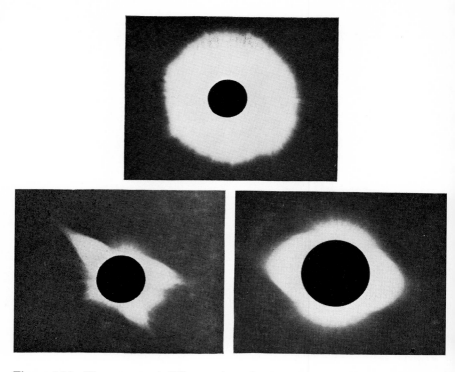

Figure 4.10. The corona at different times in the solar cycle. (*A*) Top photograph taken at time of solar maximum. (*B*) Bottom left photograph taken at solar intermediate time. (*C*) Bottom right photograph taken at solar minimum. Top and bottom left from G. Van Biesbroeck, Yerkes Observatory, and lower right from A. Wallenquist, Uppsala.

streamers are elongated at the sun's equator. They seem to stream out at the poles like iron filings from a bar magnet, possibly along the sun's magnetic field lines.

Strange emission lines from the corona were attributed to a new element, coronium, until the Swedish physicist Edlen 25 years ago showed that they were from the well-known elements such as calcium and iron. He showed that they come from metastable excited states of ionized atoms which have lifetimes long compared to the usual 10^{-8} sec. The long lifetimes are possible in the low ion densities of the corona where the times between collisions are very long. Because of the long column of low density gas of the corona, however, enough atoms are excited to be observed from the earth.

Also these lines are strongly ionized. More highly ionized than has been possible in the laboratory. Among others, the lines from Fe xxvi and Fe xxvii in the X-ray region from 1.3 to 2.0 A have been identified[8]; therefore, the temperature of the corona is very high. Several methods have now established that this temperature is about 10^6 °K. These include X-ray measurements and radio observations. Differences of a factor of 2 in the different kinds of measurements have not yet been satisfactorily explained.

4.8 THE MAGNETIC FIELD

The magnetic field of the sun plays a dominant role in the sun's atmosphere—the photosphere, the chromosphere and the corona. Excellent reviews[9,10] give details of its measurement and its influence. The magnetic field is probably responsible for the 11 year sunspot cycle and the latitude variation over the solar cycle. It is responsible for the bipolar sunspots and the dominance of preceding over following spots. The cyclic variation of the streamers in the corona is dependent on the sun's magnetic field. Although the magnetic field is very important for understanding the sun, it was not measured until 1908 when G. E. Hale[11] used the Zeeman Effect to measure magnetic fields in sunspots. All optical measurements of the magnetic field, even today, are dependent on this effect.

An electron attached to an atom in the presence of a magnetic field B behaves like a spinning bar magnet. In the normal Zeeman effect the magnetic moment precesses about the magnetic field direction with a frequency,

$$\nu_L = \frac{eB}{4\pi m_e c},\qquad (4.10)$$

called the Larmor frequency. e is the electron charge, m_e is the mass of the electron and c is the velocity of light. An energy shift $h\nu$ is added to or subtracted from the energy levels of the atom. h is Planck's constant. The frequency ν of the emitted light is then the frequency in the absence of the magnetic field ν_0 plus or minus ν_L. Along the direction of the magnetic field the two displaced frequencies are circularly polarized, one right and the other left. In a direction perpendicular to the magnetic field there are three frequencies. The center frequency, unchanged, is linearly polarized with the magnetic field direction in the plane of polarization. The two displaced frequencies are linearly polarized with the magnetic field direction perpendicular to the plane of polarization. Usually the component of magnetic field along the direction of the line of sight—the radial component is measured.

The magnetic fields in sunspots are often a few thousand gauss. It can be seen from Eq. 4.10 that a magnetic field of 3,000 gauss is required to give a displacement of 0.03 A, about 1/3 of the half-width of a Fraunhofer line. Hale and his coworkers were able to observe shifts of this magnitude with a quartz quarter-wave plate and a Nicol prism analyzer. They observed the magnetic fields in sunspots continuously from 1917 to 1924.[12] However, they were not able to measure the low general magnetic field of the sun of about 1 gauss.

The major magnetic field measurements were continued at the solar observatory at Mt. Wilson and in 1952 H. W. Babcock[13] succeeded in measuring the sun's general magnetic field. Spurious signals from the rotating analyzers had thwarted previous investigators. Babcock changed to an electrically excited crystal of ammonium dihydrogen phosphate that becomes birefringent on the application of a voltage. It was used as an oscillating circular analyzer in combination with a large plane grating that had a linear dispersion of 1 A/cm. Using magnetographs of this kind, Zeeman shifts of less than 5×10^{-5} A were measured. With a good signal to noise ratio and good optical efficiencies, angular resolutions from 5 to 20 sec were obtained with a total scan time over the sun of 1 hour. With similar apparatus velocities of 50 m/sec could also be measured.

In 1952, the magnetic field pointed outward at the sun's north pole[14] —outward at the north pole is defined as positive. And the south pole was negative. The total magnetic flux from the sun's pole was about 8×10^{21} Maxwells. In 1957 the sun's south pole reversed and became positive and in 1958 the north pole became negative. The reversal occurred at the time of the maximum in the solar cycle.

Near the equator the magnetic fields are usually bipolar, the leading pole has the sign of the pole field and the following pole the opposite sign. The bipolar magnetic regions migrate toward the equator during the solar cycle to form the butterfly diagram of Maunder[15] given in Fig. 4.11. The variation of the critical latitude with time is given by Sporer's law

$$\sin \phi_c = \frac{\pm 1.5}{y + 3} \qquad (4.11)$$

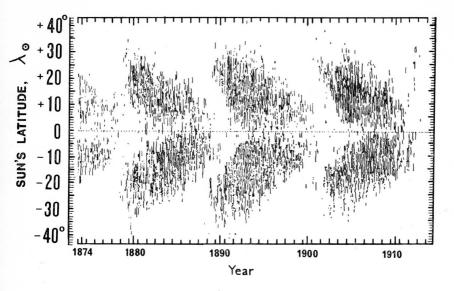

Figure 4.11. (*A*) Top: Maunder butterfly diagram. Shows the distribution of sunspot centers in latitude plotted over three solar cycles. From E. W. Maunder,[15] 1922. (*B*) Bottom: a plot of $\sin \phi_c = \dfrac{\pm 1.5}{y+3}$ for two solar cycles. The curved lines represent the latitude of the emergence of bi-polar magnetic regions or sunspot groups. From Babcock,[9] 1963.

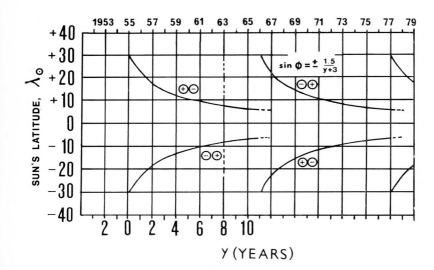

y is the elapsed time in years. Ca flocculi, faculae, spots, flares, and active prominences appear in the early stages of the bipolar magnetic region.

The magnetic field lines in the sun are frozen into the sun's plasma. They move only because the plasma is moving. In a stationary medium, $v = 0$, the diffusion time may be estimated from

$$t_d = 4\pi\sigma_c\, l^2 \tag{4.12}$$

where σ_c is the conductivity and l is the characteristic dimension. In the solar interior the conductivity might be about 10^{-4} e.m.u. and the characteristic dimension perhaps $0.1 R_\odot$, 10^{10} cm. The resultant diffusion time is then 10^{10} years, even longer than the age of the solar system. In the chromosphere with $\sigma_c = 3 \times 10^{-8}$ e.m.u. and $l = 3000$ km the diffusion time is 1800 years. Even in the chromosphere the time is much too long for the magnetic field lines to diffuse through the plasma.

Babcock[16] explains the changes in the sun's normal magnetic field in the following way. Initially 3 years before the maximum of the sunspot cycle the pole magnetic field lines are nearly in the meridian plane. The north and south poles are joined by the submerged magnetic field lines at a depth of $0.1 R_\odot$. The external connection is made at great distances from the sun. The more rapid rotation of the sun at the equator than at the poles wraps the lines around the sun near the equator. After three years the equator has gained 2.1 turns on the latitude at 30° and the magnetic field at 30° is amplified by 46 times. A critical value of the magnetic field is reached and instabilities bring flux loops to the surface as seen in Fig. 4.12. The critical value of latitude decreases with time.

The bipolar magnetic regions usually expand until they lose their identity by merging with other regions. Most of the merging is to the east or west with some migration toward the equator. In the later case the preceding part of the bipolar field merges with the preceding magnetic pole in the opposite hemisphere. The following part of the bipolar region then migrates toward the sun's pole. In this way the magnetic field of the sun is reversed.

4.9 SUNSPOTS

The sunspots are dark spots on the sun that vary in size up to 100,000 km in diameter. The dark center, called the umbra, is surrounded by a grey region called the penumbra. The umbra and penumbra are

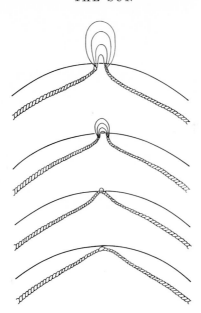

Figure 4.12. Sketch of flux loops emerging from surface of the sun. The evolution of the growth of the bipolar magnetic field is shown from the bottom to the top of the sketch.

clearly shown in photographs of sunspots in the light of Balmer $H\alpha$, $H\alpha + 0.65$ A and $H\alpha + 1.5$ A of Fig. 4.13.[17] The streaks of white and black in the light of Balmer $H\alpha$ seem to follow the magnetic field lines that are expected to exist around a bipolar magnetic field. These lines are still visible when the wavelength is shifted away by 0.65 A but disappear after the shift of 1.5 A. Then the sunspot looks much as if the photograph were taken in white light.

A spot or spot group may last from a few hours to a few days or months. The sunspot intensity has been measured since about 1650. The Wolf relative sunspot number is most often used as a measure of sunspot activity.

$$R = k_{ob} (10g + f). \tag{4.13}$$

f is the number of individual sunspots and g is the number of sunspot groups. k_{ob} is a number assigned to each observer or observatory to correct for different efficiencies and criteria for sunspots. A 27 day mean over a solar rotation is usually taken. The variation[18] in the sunspot number since 1700 is plotted in Fig. 4.14. The average period

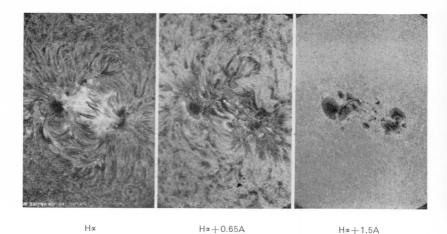

Hα   Hα+0.65A   Hα+1.5A

Figure 4.13. Photographs of sunspots taken in the light of hydrogen-Balmer α, Hα, Hα + 0.65 A and Hα + 1.5 A. From Earle Mayfield, Solar Observatory, Aerospace Corporation, El Segundo, California, 1967.

is 11 years and the maxima vary from cycle to cycle. The largest activity ever measured occurred in 1958 at the maximum of the last solar cycle. The latitude of the sunspots and the variation of the latitude in time was previously given by the butterfly diagram in Fig. 4.11.

A photograph[19] of the progression of a sunspot group across the face of the sun is shown in Fig. 4.15. The photographs were taken on the days from August 22 through September 2, 1966. A large flare occurred on August 28 and secondary flares occurred on August 31 and September 2.

For a sunspot to be in equilibrium with its surroundings its gas pressure plus magnetic pressure must equal the pressure outside in the photosphere

$$p_{\text{sunspot}} + \frac{B^2}{8\pi} = p_{\text{photosphere}} \tag{4.14}$$

Since part of the pressure in the umbra of the sunspot is furnished by the magnetic field, it is colder than the surrounding photosphere. For an umbra temperature of 4600° K and a photosphere temperature of 5750° K, Eq. 4.14 gives a magnetic field of about 100 gauss in agreement with the observed fields in sunspots.

Sunspots are often accompanied by brighter regions called faculae.

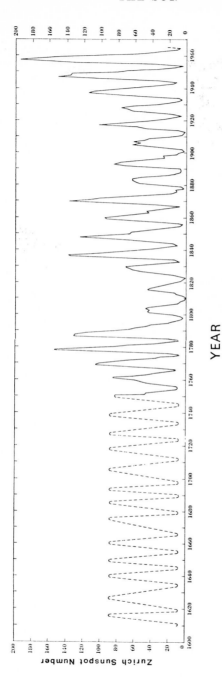

Figure 4.14. Sunspot numbers from 1610 to 1964. The times of maxima and minima are known, but the peak and valley heights were estimated from 1610 to 1749. They were measured from 1749 to 1964. Data prior to 1952 from Waldmeier as quoted by Kiepenheuer, "Solar Activity," in *The Sun*, G. P. Kuiper (ed.), University of Chicago Press, 1953. After 1952 the Zurich sunspot number data, as published in Sky and Telescope, were used.

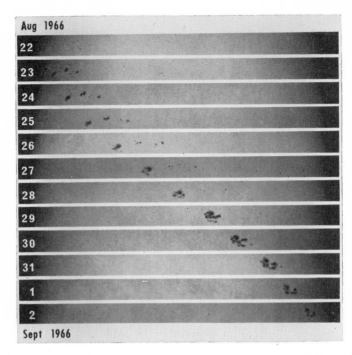

Figure 4.15. Progression of a sunspot across the face of the sun for twelve days during August and September, 1966. From Earle Mayfield, Solar Observatory, Aerospace Corporation, El Segundo, California, 1967.

Faculae seen in white light near the limb are named photospheric faculae, and those observed in the light of Ca K or Balmer Hα, chromospheric faculae. Flocculi, faculae, spots, flares, and active prominences appear in the early stages of the bipolar magnetic region.

4.10 THE QUIET SUN

Radiation is emitted from the quiet sun over the entire electromagnetic spectrum up to energies of about 5 keV (wavelength of 2 A). However, only those radiations that can penetrate the earth's atmosphere are observed on the surface of the earth. The flux of electromagnetic energy from the quiet sun is given in Fig. 4.16. Most of its energy is carried in the visible region since the sun acts as a black body with a temperature of about 5,800° K. This visible radiation passes easily through the earth's atmosphere. In the ultraviolet the radiation is cut

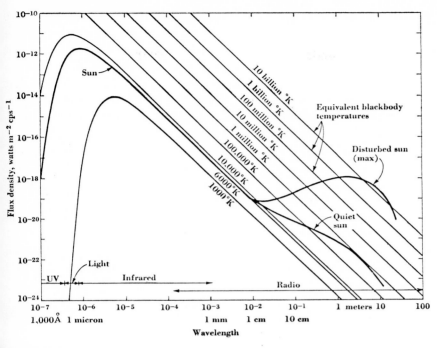

Figure 4.16. A plot of the sun's radiated flux versus wavelength. Curves are plotted for black body radiation at temperatures from 1,000 to 10 billion °K. The observed distributions for the quiet and disturbed sun are shown. These curves deviate from black body radiation for wavelengths longer than 1 cm. From Kraus,[23] 1966.

off at about 3,000 A by the ozone in the earth's atmosphere and at about 1.38 microns by water vapor absorption in the infrared (see Fig. 4.6). The UV radiation and X-rays at shorter wavelengths are completely absorbed by the earth's atmosphere so they must be observed from rockets or satellites above the atmosphere. At longer wavelengths a few discrete holes exist in the water molecular absorption so that 1 to 3 mm waves can penetrate to radio telescopes. Radio waves with wavelengths between 8 mm and 30 m penetrate to the earth's surface. Wavelengths longer than 30 m are reflected by the ionosphere so they have difficulty reaching the earth.

The visible light energy originates mostly in the photosphere, the Balmer Hα at 6563 A in the lower chromosphere and the Ca K at 3934 A in the upper chromosphere.

The H Lyman α flux at 10 eV (1216 A) at the earth is about 3×10^{11}

photons/cm²-sec and the He II flux at 40.7 eV (303.8 A) is about 3×10^9 photons/cm²-sec, down by a factor of 100.[20] The X-ray lines originate in the corona. In one such spectrum[21] (Fig. 4.17a), 54 X-ray lines between 8 and 25 A were observed. See Table 4.3 for the identification and intensity of the lines. Lines of Fe XX, Ne X and Mg XII at temperatures of over 1 million °K have been identified. During the sunspot cycle minimum in 1965 a flux of 11 photon/keV-sec-cm² was observed at 2.5 keV and of 0.5 photon/keV-sec-cm² at 3.5 keV.[22] No signal above background was seen at higher energies. The photon energy distribution was very steep, falling off a factor of 20 in 1 keV.

Radio wave radiation from the sun has been monitored almost continuously since its discovery during the war by radar receivers in England, the United States and Germany. The sun is observed at wavelengths from 1 mm to 10 m. The power received by a detector on the earth at radio wavelengths as calculated from Planck's radiation law is

$$P = \frac{2\pi k T}{\lambda^2} \frac{4\pi R_\odot{}^2}{4\pi r_{se}^2} \frac{\text{watt}}{\text{m}^2\text{–c/s}} \tag{4.15}$$

k is the Boltzman constant, 1.38×10^{-23} Joule/°K, T is the absolute temperature in °K and λ is the wavelength in meters. Multiplying by $4\pi R_\odot{}^2$ gives the total power radiated from the sun per unit frequency bandwidth and dividing by $4\pi r_{se}^2$ gives the power per unit area per unit frequency bandwidth at the earth (energy flux per c/s). For the lower corona at a temperature of 10^6 °K and at a wavelength of 1 m, the calculated power at the earth is 18.5×10^{-22} watt/m²–c/s.

The calculated curves for black body temperatures from 1000° K to 10 billion °K are given in Fig. 4.16.[23] The curves for the black body temperatures are labeled. The heavy lines give the observed curves for the quiet sun and for a typical disturbed sun. At wavelengths longer than 1 cm the experimental curves deviate orders of magnitude from a single constant temperature black body curve.

The diameter of the sun is not the same at different wavelengths. For wavelengths shorter than 1 cm the diameter is nearly the same as for visible light. At longer wavelengths, however, there is a limb brightening and the diameter appears much larger. At meter wavelengths the sun extends to several optical $R_\odot$. Radio signals from discrete radio stars are scattered by the solar corona out to 100 solar

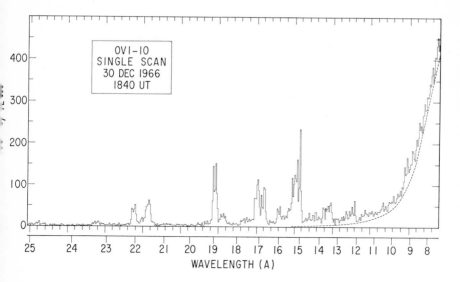

Figure 4.17. X-ray spectra from the sun measured with a Bragg Spectrometer Crystal on the Satellite OVI-10 (1966-111B). (*A*) Upper graph from the quiet sun. (*B*) Lower graph from an Importance 3B flare. The identification of the lines is given in Table 4.3. From Rugge and Walker,[21] 1968.

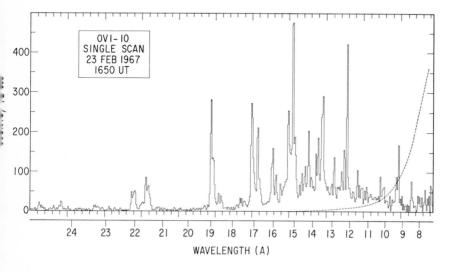

Table 4.3[21] Identification and intensities of X-ray lines
in Figs, 4.17a, b.

Observed wavelength	Predicted wavelength	Ion and transition	Line intensity 10^{-5} ergs/cm²–sec		Remarks
			22 Dec. 23:20 UT	13 Feb. 18:10–18:34 UT	
24.79	24.781	N VII	$1s^2\,S_{1/2}-2p^2P_{1/2,3/2}$	24	34	
24.58	2×12.26	Fe XVII	$2p^6\,^1S_0-2p^54d\,^3D_1$	†	—	Second order
24.28	2×12.13	Ne X	$1s^2\,S_{1/2}-2P_{1/2,3/2}$	†	—	Second order
24.22	2×12.12	Fe XVII	$2p^6\,^1S_0-2p^54d\,^1P_1$	†	—	
$23.9\ \pm.2$	—		Continuum	27.5/A	111/A	
23.29	23.28	O III	$1s^2\,2s^2\,2p^2-1s\,2s^2\,2p^3$	40	72	Identification bas calculations by He
$22.65\pm.35$	—		Continuum	45.5/A	163/A	
23.12	2×11.558	Ne IX	$1s^2\,^1S_0-1s3p\,^1P_1$	†	—	Second order
22.08	22.09	O VI	$1s^2\,2p-1s2p^2$	151	214	Identification bas calculations by H
21.80	21.80	O VII	$1s^2\,^1S_0-1s2p^3\,P_1$	40	49	
21.60	21.60	O VII	$1s^2\,^1S_0-1s2p\,^1P_1$	141	265	
$21.45\pm.30$	—		Continuum	30/A	114/A	
20.90	20.910	N VII	$1s^2\,S_{1/2}-3p\,^2P_{1/2,3/2}$	—	—	Very weak
$20.40\pm.30$	—		Continuum	18/A	54/A	
$19.75\pm.30$	—		Continuum	25.5/A	62/A	
18.96	18.969	O VIII	$1s^2\,S_{1/2}-2p\,^2P_{1/2,3/2}$	174	420	
18.64	18.627	O VII	$1s^2\,^1S_0-1s3p\,^1P_1$	38	52	
18.35	$2\times\,9.17$	Mg XI	$1s^2\,^1S_0-1s2p_1\,^{1,3}P_1$	†	—	
$18.05\pm.30$	—	—	Continuum	15/A	60/A	
17.78	17.77	O VII	$1s^2\,^1S_0-1s4p\,^1P_1$	7	*	The high density
	17.67	Fe XVI	$2p^6nl-2p^5nl\,n^1l^1$			Fe XVI lines make separation difficul
17.64	17.61	Fe XVI	$2p^6nl-2p^5nl\,n^1l^1$	8	*	The high density
	17.57	Fe XVI	$2p^6nl-2p^5nl\,n^1l^1$			Fe XVI lines make separation difficul
	17.491	Fe XVI	$2p^63s\,^2S_{1/2}-2p^53s^2\,^2P_{1/2}$			
17.42	17.39	Fe XVI	$2p^6nl-2p^5nl\,n^1l^1$	7	*	
	17.33	Fe XVI	$2p^6nl-2p^5nl\,n^1l^1$			
17.20	17.206	Fe XVI	$2p^63s\,^2S_{1/2}-2p^53s^2\,^2P_{1/2}$	6	*	
	17.16	Fe XVI	$2p^6nl-2p^5nl\,n^1l^1$			
	17.12	Fe XVI	$2p^6nl-2p^5nl\,n^1l^1$			
17.06	17.051	Fe XVII	$2p^6\,^1S_0-2p^53s\,^3P_1$	118	421	
16.77	16.774	Fe XVII	$2p^6\,^1S_0-2p^53s\,^1P_1$	74	256	
$16.4\ \pm.2$	—		Continuum	‡	<92/A	Upper limit, flux m due to unresolved
16.30				4.5	35	
16.15	16.17	Fe XVIII	$2s\,2p^6-2s\,2p^5\,3s$	5	*	
	16.08	Fe XVIII	$2p^5-2p^4\,3s$			
16.01	16.01	Fe XVIII	$2p^5\,^2P-2p^4(^3P)3s\,^4P$	23	166	
	16.006	O VIII	$1s\,^2S_{1/2}-3p\,^2P_{1/2,3/2}$			
15.88	15.88	Fe XVIII	$2p^5\,^2P-2p^4(^3P)3s\,^4P$	14.5	61.5	
	15.82	Fe XVIII	$2p^5-2p^4\,3s$			
	15.76	Fe XVIII	$2p^5-2p^4\,3s$			
15.61	15.62	Fe XVIII	$2p^5\,^2P-2p^4(^1D)3s\,^2D$	7.5	56.5	
15.45	15.453	Fe XVII	$2p^6\,^1S_0-2p^5\,3d\,^3P_1$	21	} 271	
15.25	15.261	Fe XVII	$2p^6\,^1S_0-2p^5\,3d\,^3D_1$	56.5		
	15.17	Fe XIX	$2p^4-2p^3\,3s$			
	15.09	Fe XIX	$2p^4-2p^3\,3s$			
	15.06	Fe XIX	$2p^4-2p^3\,3s$			
15.01	15.012	Fe XVII	$2p^6\,^1S_0-2p^5\,3d\,^1P_1$	92.5	314	
	14.92	Fe XIX	$2p^4-2p^3\,3s$			
	14.79	Fe XVIII	$2p^5-2p^4\,3d$			
14.68	14.69	Fe XVIII	$2p^5-2p^4\,3d$	4	41	
	14.634	O VIII	$1s\,^2S_{1/2}-6p\,^2P_{1/2,3/2}$			
	14.59	Fe XVIII	$2p^5-2p^4\,3d$			
14.56	14.57	Ne III	$1s^2\,2s^2\,2p^4-1s\,2s^2\,2p^5$	7	61	Ne III identifi based on calculatio
	14.54	Fe XVIII	$2p^5-2p^4\,3d$			House
	14.40	Fe XVIII	$2p^5\,^2P-2p^4(^3P)3d\,^2D$			
14.39				8	66	
	14.37	Ni XVIII	$2p^6\,3s\,^2S-2p^5\,3s^2\,^2P_{3/2}$			
	14.29	Fe XVIII	$2p^5-2p^4\,3d$			

Observed wavelength	Predicted wavelength	Ion and transition	Line intensity 10^{-5} ergs/cm^2–sec 22 Dec. 23:20 UT	13 Feb. 18:10–18:34 UT	Remarks
7			14.5	114	
	14.25	Fe XVIII	$2p^5\,^2P - 2p^4(^1D)3d\,^2D$		
)	14.10	Ni XVIII	$2p^6\,3s\,^2S - 2p^5\,3s^2\,^2P_{1/2}$	6.0 }	56
2	14.03	Ni XIX	$2p^6\,^1S_0 - 2p^5\,3s\,^3P_1$		
2	13.887	Fe XVII	$2p^6\,^1S_0 - 2p^5\,3p\,^3P_1$		
	13.83	Ne VII	$1s^2\,2p^2 - 1s\,2p^3$		116
2	13.82	Fe XVII	$2p^6\,^1S_0 - 2p^5\,3p\,^1P_1$	10.5	
	13.77	Ni XIX	$2p^6\,^1S_0 - 2p^5\,3s\,^1P_1$		
4	13.72	Fe XX	$2s^2\,2p^3 - 2s^2\,2p^2\,3s$	11.0	Ne VIII identification based on calculations by House
	13.66	Ne VIII	$1s^2\,2p - 1s\,2p^2$		
	13.66	Fe XX	$2s^2\,2p^3 - 2s^2\,2p^2\,3s$		173
	13.55	Ne IX	$1s^2\,^1S_0 - 1s\,2p\,^3P_1$		
8	13.51	Fe XIX	$2p^4 - 2p^3\,3d$	21.0	
	13.46	Fe XIX	$2p^4 - 2p^3\,3d$		
	13.44	Ne IX	$1s^2\,^1S_0 - 1s\,2p\,^1P_1$		
	13.40	Fe XIX	$2p^4 - 2p^3\,3d$		
	13.31	Fe XIX	$2p^4 - 2p^3\,3d$		
4	12.98	Fe XX	$2p^3 - 2p^2\,3d$	†	
	12.93	Fe XX	$2p^3 - 2p^2\,3d$		102
5	12.83	Fe XX	$2p^3 - 2p^2\,3d$	5	
	12.80	Ni XIX	$2p^6\,^1S_0 - 2p^5\,3d\,^3P_1$		
)	12.64	Ni XIX	$2p^6\,^1S_0 - 2p^5\,3d\,^3D_1$	5	‡
	12.60	Fe XXI	$2p^2 - 2p\,3d$		
5	12.42	Ni XIX	$2p^6\,^1S_0 - 2p^5\,3d\,^1P_1$	4.5	50
6	12.26	Fe XVII	$2p^6\,^1S_0 - 2p^5\,4d\,^3P_1$	8	83
4	12.134	Ne X	$1s\,^2S_{1/2} - 2p\,^2P_{1/2,3/2}$	13	75
	12.12	Fe XVII	$2p^6\,^1S_0 - 2p^5\,4d\,^1P_1$		
6		Ne (?)	$1s - 3p$ Satellite (?)	‡	38
	11.59	Ni XIX	$2s^2\,2p^6\,^1S_0 - 2s\,2p^6\,3p\,^3P_1$		
7	11.558	Ne IX	$1s^2\,^1S_0 - 1s\,3p\,^1P_1$	5	40
1	11.53	Ni XIX	$2s^2\,2p^6\,^1S_0 - 2s\,2p^6\,3p\,^1P_1$	†	34
8	11.16	Na IX	$1s^2\,2p - 1s\,2p^2$	‡	‡	Na IX identification based on calculations by House
5	11.08	Na X	$1s^2\,^1S_0 - 1s\,2p\,^3P_1$	‡	26
	11.00	Na X	$1s^2\,^1S_0 - 1s\,2p\,^1P_1$		
2				1	‡
1					27
1	10.66	Fe XXIV	$2p - 3d$	2	
0				2	‡
5	10.239	Ne X	$1s\,^2S_{1/2} - 3p\,^2P_{1/2,3/2}$	2	‡
5	10.10	Ni XIX	$2p^6\,^1S_0 - 2p^5\,4d\,^1P_1$	1.0	25
	10.025	Na XI	$1s\,^2S_{1/2} - 2p\,^2P_{1/2,3/2}$		
	9.97	Ni XIX	$2p^6\,^1S_0 - 2p^5\,4d\,^3P_1$		
5	9.708	Ne X	$1s\,^2S_{1/2} - 4p\,^2P_{1/2,3/2}$	3	16
3	9.48	Ne X	$1s\,^2S_{1/2} - 5p\,^2P_{1/2,3/2}$	4	‡
	9.39	Mg XI	Satellite (2)		
	9.362	Ne X	$1s\,^2S_{1/2} - 6p\,^2P_{1/2,3/2}$		
3	9.29	Mg X	$1s^2\,2p - 1s\,2p^2$	4		Mg X identification based on calculations by House
	9.23	Mg XI	$1s^2\,^1S_0 - 1s\,2p\,^3P_1$		94
8	9.16	Mg XI	$1s^2\,^1S_0 - 1s\,2p\,^1P_1$	10	
5		Mg (?)	Satellite (?)	†	
4	8.459	Na XI	$1s\,^2S_{1/2} - 3p\,^2P_{1/2,3/2}$	†	31
	8.421	Mg XII	$1s\,^2S_{1/2} - 2p\,^2P_{1/2,3/2}$		

†Line not observed in this spectrum
‡Line not resolved in this spectrum, but probably present
◦Continuum obscured by lines

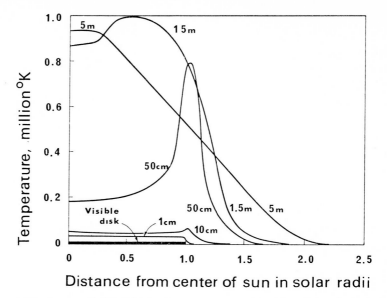

Figure 4.18. Apparent blackbody temperature across the disk of the sun at different radio wavelengths.   From Smerd,[24] 1950.

radii.   It is customary to put the measured power into Eq. 4.15 and solve for the temperature.   The power from the sun and other sources is then measured in terms of an equivalent temperature.   Equivalent theoretical temperatures[24] at different wavelengths versus distance out from the center of the sun are plotted in Fig. 4.18.   These curves suggest that the shortest wavelengths originate close to the photosphere and that the longer wavelengths are produced at successively higher levels in the chromosphere and corona.   The radiation at any wavelength comes from a thin layer just below the height whose critical density is given in Eq. 3.5

$$n_0 = \frac{1.12 \times 10^{13}}{\lambda_0^2} \frac{\text{electrons}}{\text{cm}^3} \qquad (4.16)$$

At wavelengths longer than the critical wavelength at a particular height the waves rapidly lose energy to the surrounding plasma and are not propagated outward.

The radio wave energy flux varies over the solar cycle.   At 10.7 cm it varies from about 65×10^{-22} W/m²–c/s at solar minimum to about 250×10^{-22} W/m²–c/s at solar maximum.   The variation from 1947 to 1963 is shown in Fig. 2.7.

The sun also continuously emits a solar wind of low energy electrons and protons. In 1964 and 1965 during the sunspot minimum these particles had an average velocity of 320 km/sec.[25] The proton average energy was about 1 keV and the electron average energy about 1 eV. The velocities varied between 280 and 750 km/sec. The average velocity direction was 1.5 deg east of the sun and varied over 15 deg. Temperatures perpendicular to the flow velocity of the protons varied downward from 10^5 °K and were typically 10^4 °K when the flow velocity was low. α-particles with abundances of about 5% of the protons and of the same velocities as the protons were also measured.

4.11 SOLAR FLARES

A solar flare is observed as the temporary occurrence of strong emission light inside normally dark Fraunhofer lines. The spots of bright light in solar flares are often monitored in Balmer Hα. The burst of light rises in a few minutes and decays away in a few tens of minutes. Flares are classified by area and brightness into importance 1$^-$, 1, 2, 3, and 3$^+$, running from the smallest to the largest. The small flares are much more frequent than the large ones. In 1960 the ratios of the numbers of flares of importance 1 to 2 to 3 were 91 to 8 to 1%. Importance 1 flares occur in the vicinity of sunspots and often near a multipolar magnetic field group, a γ group. Photographs of the development of a flare are shown in Fig. 4.19. Limb flares can be photographed with a coronagraph that blocks out the light from the disk of the sun. Examples of surge prominences seen with a coronagraph are shown in Fig. 4.20. Some of these go out as far as 100,000 km with velocities of 500 km/sec.

4.11.1 X-Rays

At times of solar flares the X-ray intensities increase much more than the visible or ultra-violet. R. W. Kreplin, T. A. Chubb and Herbert Friedman[26] measured the X-ray fluxes from the sun with a Be window ion chamber sensitive from 2 to 8 A (6 keV to 1.5 keV) from June 22 to November 1, 1960 on the SR-1 satellite near the time of the maximum of the solar cycle. They found that the normal X-ray fluxes during this time were less than the threshold of 0.6×10^{-3} erg/cm²-sec (2×10^5 photons/cm²-sec). More than 100 cases were observed where the flux exceeded the threshold. In these cases one or more distinct solar events

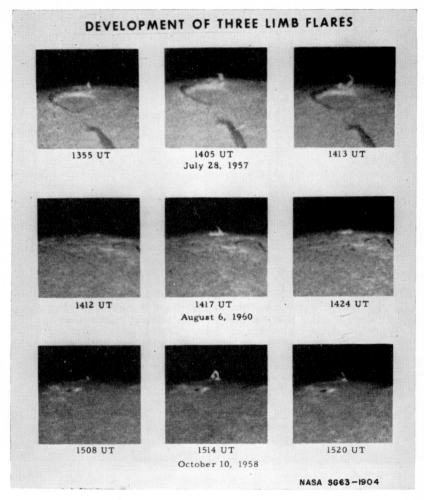

Figure 4.19. Photographs of the development of three solar limb flares. From Henry J. Smith, "Observations on Flares," AAS-NASA Symposium on the Physics of Solar Flares, NASA SP-50. Scientific and Technical Information Division, NASA, Washington, D.C., 1964.

were visible on the sun. And shortwave fadeout and other ionospheric effects were noticeable on the earth.

A typical example of X-ray increases that occur during solar flares is shown in Fig. 4.17b and Table 4.3.[21] During the 3B flare on Feb. 13 1967 the intensities of the X-ray lines increased by a factor of 10 at X-ray wavelengths of 8 to 15 A.

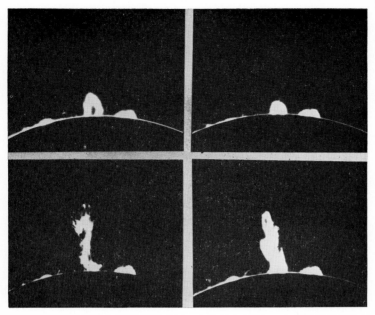

Figure 4.20. Photograph of a limb flare and eruption on June 14, 1956 with an Hα filter at Sacramento Peak. From Harold Zirin, "Spectroscopic Observations," AAS-NASA Symposium on the Physics of Solar Flares, NASA SP-50, Scientific and Technical Information Division, NASA, Washington, D.C., 1964.

The X-ray fluxes during the minimum of the solar cycle were measured by R. L. Arnoldy, S. R. Kane and J. R. Winckler[27] on OGO-A and OGO-B spacecraft. In over 6,000 hours of observation outside the earth's magnetosphere, over 30 solar flare X-ray bursts were detected and correlations were made with radio and energetic particle emission. For a given flare, the rise times, decay times and total duration of the radio and X-ray bursts were similar. Rise times were usually less than a minute and decays were exponential with time constants between 1 and 10 minutes. An example of the correlation of X-rays with 3 and 10 cm waves is shown in Fig. 4.21. The rise times and average decay rates are similar. The secondary bumps in the radio wave distributions are at best barely recognizable in the X-ray data. All 3 and 10 cm radio bursts with intensities greater than 80×10^{-22} W/m²–c/s were accompanied by X-ray events with energy fluxes greater than 3×10^{-7} ergs/cm²-sec (20 photons/cm²-sec). All but three of the X-ray events were accompanied by ionospheric disturbances. In many of the large

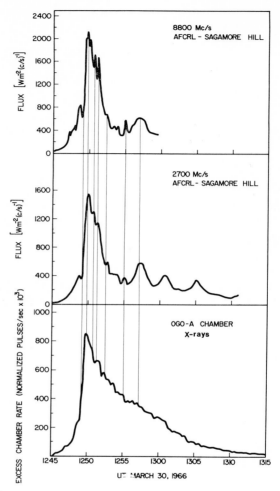

Figure 4.21. Comparison of Radio power at 8800 Mc/s and 2700 Mc/s with X-rays from the sun. The time variations during the flare of March 30, 1966 are shown. From Arnoldy, Kane and Winkler,[27] 1967.

flares solar protons and solar electrons were also present. However, many solar proton events occurred without a detectable burst of energetic X-rays.

X-rays with energies greater than 20 keV were measured in three importance 2 flares in 1959.[28] The energy distribution for the flare of September 1 is shown in Fig. 4.22. The triangles are measurements from the direction of the sun, the squares from the direction of the earth,

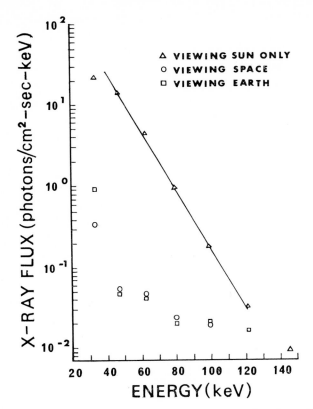

Figure 4.22. X-ray energy distribution from a class 2 solar flare, September 1, 1959. Data obtained by a scintillation detector on a rocket. From Chubb, Kreplin, and Friedman,[28] 1966.

and the circles from other directions. The data could be explained by bremsstrahlung in a 160 million degree plasma on the surface of the sun.

4.11.2 Radio wave bursts

In 1942 a chain of British radar stations discovered very strong noise signals from the sun at wavelengths of 4–6 m that lasted for a couple of days. Hey[29] investigated and found that they were caused by a large solar flare on the sun. Since that early work, radio burst research has contributed significantly to our information about the sun. The solar radio bursts up to 1963 are the subject of an excellent review article by J. P. Wild, S. F. Smerd and A. A. Weiss.[30]

Although the radio wave energy is only a small fraction of the total flare energy, radio flares are spectacular, particularly at meter wavelengths. The intensity of the radio flare increases by a factor of a thousand in seconds and the apparent temperature occasionally rises by a factor of one hundred thousand above the normal temperature of 10^6 °K. The radio bursts are sensitive indicators of eruptions on the sun. Radio bursts provide information about the injection, acceleration and trapping of fast electrons in the sun's atmosphere during solar flares.

The intensity and variation of radio bursts change greatly with wavelength. The bursts are small at cm waves but increase in intensity and complication at longer wavelengths. The simple to the very complicated burst patterns are seen in Fig. 4.23. The cm radio waves probably originate rather low in the sun's atmosphere near the bottom of the corona or at the top of the chromosphere. The meter radio waves probably originate higher in the corona above the region of the solar flare explosions.

The solar radio bursts have been divided into 5 classes.[30] They are:

Type   I. Noise storm bursts.
Type II. Slow-drift bursts.
Type III. Fast-drift bursts.
Type IV. Broad band continuum radiation.
Type   V. Broad band continuum radiation at meter wavelengths.

The 5 classes of solar signals are given in Fig. 4.24.[30] The frequency is plotted vertically and time horizontally. The intensity of the signal is proportional to the exposure of the film.

Type I noise storms are a long series of bursts sometimes superimposed on continuous radiation. Intense storms may last for several days with several hundred storm bursts every hour. The bursts usually occur at frequencies below 250 mc/s. At sunspot maximum they occur about 10% of the time. The bunching of the Type I bursts can probably be best explained by plasma waves.

Type II bursts are associated with large flares: they accompany 2% of the importance 1 flares and 30% of the importance 3 flares. The bursts last about 10 minutes and occur at the rate of about 1 per 50 hours near sunspot maximum. The bursts drift from high to low

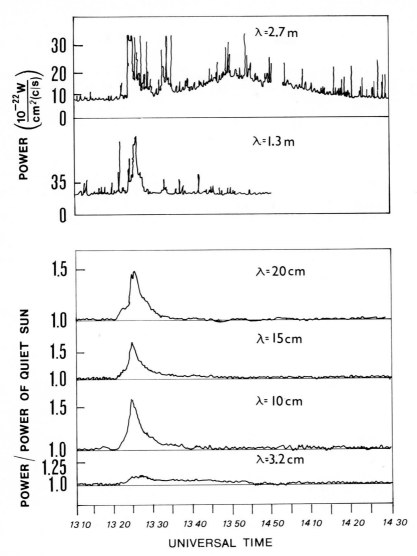

Figure 4.23. Radio-frequency power from the sun during the solar flare of May 23, 1960. Time variations are given for wavelengths from 3.2 cm to 2.7 m. Recordings of 3.2 to 20 cm were made at Berlin-Adlershof, and those of 1.3 and 2.7 m at Potsdam-Tremsdorf. From Beobachtungs Ergebnisse of Heinrich-Hertz-Institut, May, 1960.

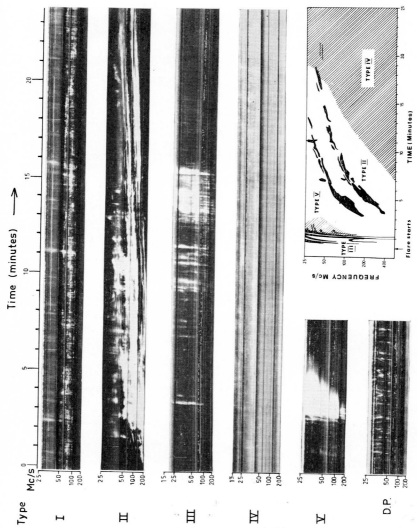

Figure 4.24. Examples of the five types of radio bursts and drifting pairs. The inset is a sketch of a complete radio burst. From Wild, Smerd and Weiss,[30] 1964.

frequencies at rates up to 1 mc/s. They usually have a narrow bandwidth and often occur with a second harmonic. The type II bursts probably originate in plasma waves.

Type III bursts are the most common of the meter bursts. The bursts last for 10 sec with a rate of occurrence greater than 3 per hour at sunspot maximum. The bursts drift rapidly from high to low frequencies at the rate of about 20 mc/sec. The fast drift in frequency and the presence of harmonics suggests that the bursts are caused by plasma waves that are excited by a disturbance that moves rapidly out through the solar corona.

Type IV bursts. This radiation is persistent, smooth, and continuous over a broad band of frequencies. The radiation lasts for tens of minutes after major solar flares. The source appears to move outward at speeds of several hundred km/sec and reach great heights. Because of the emission at great heights plasma waves are probably not the source. Rather synchrotron radiation of electrons in the magnetic fields of the corona.

Type V bursts. They occur about 10% as often as type III bursts which they usually follow. The radiation has a broad-band continuum with its maximum intensity below 150 mc/sec. Its origin is presently undetermined.

Much has been learned from the new field of radio astronomy. Much will be learned in the future, particularly during the next solar cycle that reaches its maximum in 1970. Measurements of the intensity and polarization of the radio waves at many frequencies and times with good resolution will be important. These measurements should be of great assistance in determining the mechanism for the production of solar flares.

4.11.3 Solar flare protons

During the last solar cycle with its maximum in 1959, about 10^{11} protons/cm² with energies greater than 10 MeV struck the boundary of earth's magnetic cavity. This was about 100 times larger than the total number of cosmic ray protons that arrived over the same period of time. However, the cosmic ray protons were mostly of higher energy, over 1 BeV, while the solar protons were almost all less than 1 BeV. At quiet times the magnetic field of the earth at the equator turns away all protons with momenta less than 15 BeV/c and at higher latitudes λ, all protons with momenta less than

$$pc = 15 \cos^4 \lambda \quad \text{BeV}/c. \tag{4.17}$$

This equation must be modified slightly to take into account the earth's magnetic cavity with its long tail. Except during times of magnetic storms the cut-off prohibits the entry of solar protons at latitudes below about 60 deg. Likewise the cosmic ray proton intensity increases with latitude from the equator by a factor of 4 to a plateau at about 60 deg latitude. At times of magnetic storms the minimum latitudes at which particles enter are reduced because of the ring current of particles formed around the earth. The ring current reduces the magnetic field and thus the cut-off momenta.

Very energetic solar protons were first detected by ionization chamber detectors on the surface of the earth. Beginning about 1949 the more sensitive neutron detectors on the earth monitored the secondary neutrons produced in the atmosphere by the high energy protons. The variations and fluctuations in the cosmic rays were measured at various geographical locations and times of the solar cycle. The detectors were sometimes placed on tops of mountains to get above as much atmosphere as possible to increase the neutron counting rates. At the time of the solar flare of November 19, 1949 the neutron counting rate increased by a factor of 6.[31] And in the intense flare of February 23, 1956 by a factor of 10.[32] This was the highest counting rate ever detected to that time and was clearly associated with a large visible solar flare on the sun and a magnetic disturbance on the earth. Only four solar proton events had been observed by particle detectors on the earth prior to 1954. However, 10 such events were observed between 1954 and 1961 over the peak of the last solar cycle. The increased number was due to the more sensitive detection methods.

The long column of atmosphere above the neutron monitors located at sea level is 1,000 gm/cm² thick. Consequently, only the highest energy protons, those with energies greater than about 1 BeV, make enough neutrons to be detected at sea level. In order to get above this absorbing atmosphere to detect the lower energy protons, detectors were flown on balloons. They typically rise to 100,000 ft under only 4 gm/cm² of atmosphere and can be reached by protons of 50 MeV. Much of the pioneer balloon work was carried out by John Winckler and his colleagues at the University of Minnesota. An example of the onset of the protons with energies greater than 80 MeV is shown for the flare

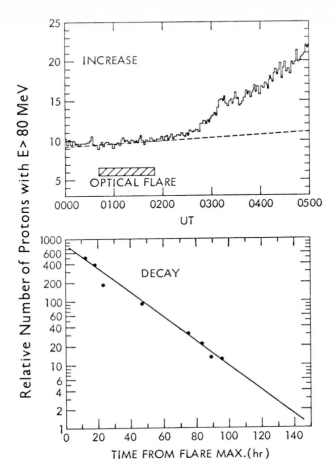

Figure 4.25. Onset and decay of protons with $E > 80$ MeV during the solar flare of September 3, 1960. The data were taken by Geiger counters and ion chambers on balloons. The upper graph shows the increase following the optical flare. UT stands for universal time. The lower graph shows the decrease with a mean lifetime of 23 hr. The flare occurred on the East limb of the sun. The dashed line is the galactic cosmic ray background. From Hoffman and Winkler,[33] 1963.

of September 3, 1960 in Fig. 4.25.[33] The rise times are longer than the proton straight line flight times from the sun to the earth. Light takes only eight minutes to travel this distance. But the protons take longer because they travel at velocities less than the velocity of light and follow spiral curved paths in the magnetic fields of interplanetary space. A 100 MeV proton that travels a straight path to the earth takes about 20 minutes and a 10 MeV proton about 1 hour.   An exponential mean decay time of 23 hr was measured.

Detectors carried on satellites can monitor solar flares continuously, particularly if they are outside the earth's magnetic cavity.   An example of the time dependence of protons at eight energies from 3.8 to 295 MeV for the proton flare of September 28, 1961 are shown in Fig. 4.26.[34]   Here the rise times are a few hours and the decay times a few days.

The rise and decay times of the flares in the last solar cycle have been summarized in a review article by William Webber.[35]   The fastest observed rise time to maximum was only 7 min. and occurred for the flare of May 4, 1960.   Other rise times varied from 30 minutes for the

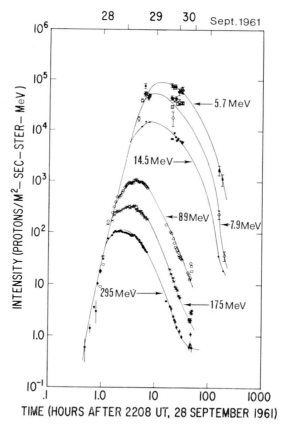

Figure 4.26. Build-up and decay of protons from the solar flare of September 28, 1961.   Data from Bryant, Cline, Desai and McDonald,[34] 1964.

flare of February 23, 1956 to 6 hours for the flare of September 3, 1960. There is a definite correlation between the rise time and the position on the solar disk where the flare was formed. The fastest rise times occur for the extreme western edge of the disk and the slowest for the extreme eastern edge.

Eugene Parker[36] and Ken McCracken[37] suggested that the solar wind carries the tightly frozen-in magnetic field in a radial direction away from the sun. But one end of the magnetic field line is tied to and rotates with the sun. The magnetic field lines curve away from the sun in an Archimedes Spiral like the spray from a rotating sprinkler. And the solar protons spiral around these magnetic field lines as they travel through interplanetary space. Their picture of the magnetic field lines is given in Fig. 4.27. The west limb is on the right. With the flare on the west limb in the position shown, the lines connected to the flare pass through the earth. If the flare occurs on the east limb, however,

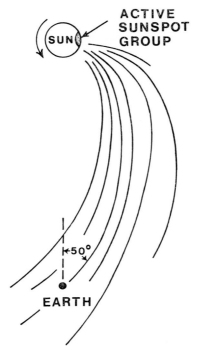

Figure 4.27. Magnetic field lines connecting the sun to the earth. A solar flare at 40 to 60 degrees west solar longitude is connected to the earth by magnetic field lines at the garden hose angle of about 50 degrees. Protons or electrons emitted from the sun at this time easily reach the earth. From McCracken,[37] 1962.

the particles must diffuse across magnetic field lines to reach the earth. The farther to the east, the more field lines to diffuse across and the longer the time for the particles to reach the earth.

During our last solar cycle, cycle 19, more than 10^5 protons/cm² reached the earth in each of an estimated 100 solar flares. In 11 of these flares the total fluxes of protons with energies greater than 10 MeV were larger than 10^9 proton/cm². The 11 events are listed in Table 4.4.[38] The numbers of protons with energies greater than 30 and

Table 4.4 Proton fluxes (particles/cm²) for the 11 largest solar proton flares in the 19th solar cycle[38]

Flare Date	Energy	> 10 MeV	> 30 MeV	> 100 MeV
Feb. 23, '56		1.8×10^9	1.0×10^9	3.5×10^8
Mar. 23, '58		2.0×10^9	2.5×10^8	1.0×10^7
July 7, '58		1.8×10^9	2.5×10^8	9.0×10^6
Aug. 26, '58		1.5×10^9	1.1×10^8	2.0×10^6
May 10, '59		5.5×10^9	9.6×10^8	8.5×10^7
July 10, '59		4.5×10^9	1.0×10^9	1.4×10^8
July 14, '59		7.5×10^9	1.3×10^9	1.0×10^8
July 16, '59		3.3×10^9	9.1×10^8	1.3×10^8
Nov. 12, '60		4.0×10^9	1.3×10^9	2.5×10^8
Nov. 15, '60		2.5×10^9	7.2×10^8	1.2×10^8
July 18, '61		1.0×10^9	3.0×10^8	4.0×10^7

100 MeV are also included. These 11 largest proton solar flares produced ten times more protons with energies greater than 10 MeV at the earth than the cosmic rays.

The solar flare proton energy distributions have been measured for many of the events. Phyllis Freier and William Webber[39] found an exponential decrease of the flux J with rigidity R.

$$J = J_0 \, e^{-(R/R_0)} \tag{4.18}$$

The rigidity is the momentum of the particle divided by its charge. It is usually given in units of BV. J_0 is the flux for $R = 0$. For a change in the rigidity of R_0 the flux changes by $1/e$. The energy distributions for 6 prominent flares are given in Fig. 4.28.[39] The flare with the most energetic protons measured directly by experiments on satellites occurred on November 15, 1960. However even more energetic protons were measured by detecting the secondary neutrons on the earth in the

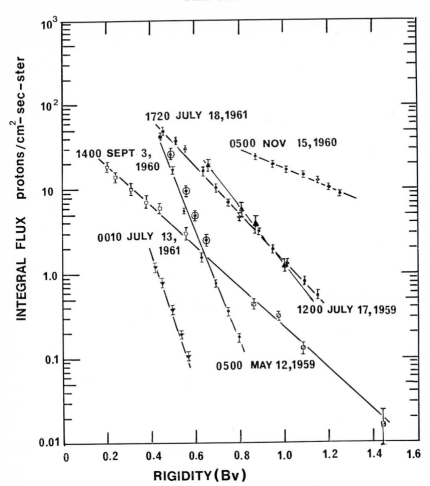

Figure 4.28. Integral proton energy distributions as a function of rigidity for six different flares. The counter data are plotted with solid symbols, the emulsion data with open symbols. From Freier and Webber,[39] 1963.

flare of February 23, 1956. A steeply rising flare like the one of May 12, 1959 gives the most low energy protons.

Both J_0 and R_0 differ from flare to flare and also within a flare. Figure 4.29 gives a plot of R_0 and J_0 for the flares of November 12 and 15, 1960.[39] J_0 climbs to a peak at the maximum of the flare and then decays away. R_0 drops off approximately exponentially during the course of the flare. The protons at the beginning of the flare are the most energetic. They become softer and softer as time goes on.

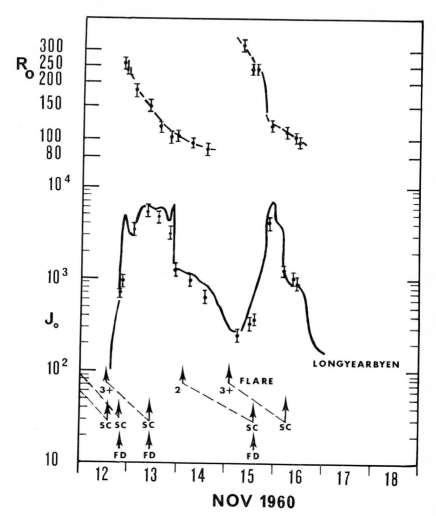

Figure 4.29. The time variations of R_0 (MV) and J_0 (protons/cm²-sec-ster) for the November 12 and 15, 1960 flares. These are the constants in the equation for the rigidity distribution

$$J = J_0 \, e^{-R/R_0} \, .$$

R is the rigidity. The continuous curve is the riometer absorption data from the Longyearbyen Station. Times of major solar flares, sudden commencements SC, and Forbush decreases FD are also indicated. From Freier and Webber,[39] 1963.

4.11.4 Heavier particles

The ratio of α-particles to protons is constant for a given solar flare but differs greatly from flare to flare. The ratio varies from 0.02 to 1. The high value of one often found in solar flares is much larger than the value of 0.07 usually found for cosmic rays.

The flux and energy spectra of helium and heavier nuclei have been measured with nuclear emulsions on rocket flights in the flares of September 3, November 12 and 15, 1960. The rigidity spectra of He, C, N, O and heavier nuclei were the same. The average ratio of the abundance of He to medium nuclei was 60.[40] The composition is similar to that of the solar atmosphere but significantly different from cosmic rays. The composition of the multiply charged nuclei seem to be the same for each event and is similar to the composition of the sun. The relative abundances of the nuclei in solar flares, the sun, cosmic rays and the universe are compared in Table 4.5.[41]

Since the proton to α-particle ratio varies from flare to flare it is not possible to obtain the hydrogen to helium abundance on the sun from the solar flare data alone. However, if the average helium to medium ratio is taken to be 62 ± 7 and the ratio of the proton to medium nuclei to be 650^2 a proton to helium ratio of 10.5 ± 4.0 is obtained.

4.11.5 Low-energy recurring particles

Magnetic storms on the earth sometimes recur with a time period of the rotation of the sun of 27 days. Julius Bartels[42] in 1932 suggested that they were caused by charged particles continuously emitted from the long-lived M regions on the sun. These magnetic storm protons have also been observed at energies of 3 to 20 MeV. The protons are contained in streams of 30 to 120 deg width at the earth and persist for many rotations of the sun.[43] Each recurrence event is immediately preceded by complex magnetic activity. The arrival times do not depend on the proton energy since the shapes of the energy distributions are similar at onset of the flare and at maximum. α-particles with fluxes down by a factor of 15 and with similar energy distributions are also found to rotate with the sun.[44]

The time of arrival of the low energy protons at the satellite IMP 3 was 50 min. earlier than for Explorer 33[45] on July 8, 1966. This delay verified that the charged particles are tied to the sun's magnetic field lines that sweep across the earth. The time that it takes for the solar

Table 4.5a   Relative abundances of nuclei normalized
to a base of 1.0 for oxygen

Element	Solar cosmic rays [*]	Sun [†]	Universal abundances [‡]	Galactic cosmic rays [§]
$_2$He	107 ± 14	?	150	48
$_3$Li	—	$<10^{-5}$	$<10^{-5}$	0.3
$_4$Be$-_5$B	<0.02	$<10^{-5}$	$<10^{-5}$	0.8
$_6$C	0.59 ± 0.07	0.6	0.26	1.8
$_7$N	0.19 ± 0.04	0.1	0.20	$\lesssim 0.8$
$_8$O	1.0	1.0	1.0	1.0
$_9$F	<0.03	0.001	$<10^{-4}$	$\lesssim 0.1$
$_{10}$Ne	0.13 ± 0.02	?	0.36	0.30
$_{11}$Na	—	0.002	0.002	0.19
$_{12}$Mg	0.043 ± 0.011	0.027	0.040	0.32
$_{13}$Al	—	0.002	0.004	0.06
$_{14}$Si	0.033 ± 0.011	0.035	0.045	0.12
$_{15}$P$-_{21}$Sc	0.057 ± 0.017	0.032‖	0.024	0.13
$_{22}$Ti$-_{28}$Ni	$\lesssim 0.02$	0.006	0.033	0.28

[*] *Biswas et al.* (1962, 1963), *Biswas and Fichtel* (1964), and this work.

[†] The uncertainty of the values in this column is probably of the order of a factor of 0.5. See *Aller* (1953) or *Goldberg et al.* (1960).

[‡] The uncertainty of the values in this column is hard to estimate, but it is probably at least a factor of 0.5 in some cases. See *Suess* and *Urey* (1956) and *Cameron* (1959).

[§] The uncertainty of the values in this column varies from 10% to about 30%. See *Waddington* (1960).

‖ A 5/2 ratio for the abundance of $_{16}$S relative to $_{18}$A was assumed, the relative abundance of $_{18}$A being unknown.

[a] Taken from Biswas, Fichtel and Guss.[41] The references, above, listed in the footnotes to the table may be found in this paper.

wind plasma with its frozen-in magnetic field to reach the earth is $t = r_{es}/v_s$, about four days. During this time the sun rotates through an angle of $\phi = t \Omega_\odot$. The $\Omega_\odot$ is the angular velocity of rotation of the sun of 13 deg/day. For a quiet time solar wind velocity v_s of 350 km/sec, ϕ is 66 deg. Consider the flare that occurred early on July 7, 1966 at a position of W 48 deg. It was necessary for the sun to rotate to a position of W 66 deg in order for the particles tied to the magnetic field lines to sweep across the earth. This required a time of 18 deg/13 deg/day or 1.4 da. in agreement with the delay of the time of arrival of particles at the earth.[45]

If the irregularities in the interplanetary magnetic field are larger than the cyclotron radii of the particles, the first adiabatic invariant $\mu = E_\perp/B$ (Eq. 1.3) is constant. Away from the sun the field decreases at least as fast as $1/r^2$. r is the distance from the sun. Then for μ to remain constant, $E_\perp$ must also decrease as fast $1/r^2$. The perpendicular velocity must decrease at least as fast as $1/r$. This velocity $v_\perp$ is $v \sin \alpha$ where v is the total velocity and α is the angle with the magnetic field, the pitch angle. The total velocity does not change so the pitch angle decreases and particles rapidly become tightly collimated along the direction of the magnetic field.

The continued presence of the protons indicates that they are continuously accelerated. It is not known whether the protons are accelerated in the immediate vicinity of the sun or in interplanetary space. Perhaps they are continuously accelerated in the disordered fields in the regions of space swept out by the rotating magnetic field lines. There is still the question of why these rotating fields of the sun, as identified by magnetic field and plasma measurements, do not *always* have the surplus fluxes of protons and α-particles.

4.11.6 Solar electrons

The first evidence for solar electrons came from ion-chamber and Geiger counter measurements of bremsstrahlung from a flux of about 10^7 electrons/cm²-sec striking the outer shell of the spacecraft Pioneer V.[46] The electrons were observed by R. Arnoldy, R. Hoffman and John Winckler during the very active time of March 27 to April 6, 1960. Numerous flares of many sizes, large loop and surge prominences and strong solar radio emissions over a wide range of frequencies occurred. A strong geomagnetic storm began on March 31 with major earth current disturbances and a complete blackout of the North Atlantic communications. The authors suggested that "These electrons in free space must be the extreme energetic part of a solar plasma stream associated with the strong geomagnetic storm on March 31 and again with the weaker event on April 3."

Next, electrons of 10 to 35 keV were seen[47] one minute before the observance of the sudden commencement on the earth's surface of a magnetic storm on September 30, 1961. The flux reached a peak of 3×10^6 electrons/ cm²-sec-ster, lasted for about 10 minutes and then disappeared.

Many additional cases[48,49,50] of solar electrons now support the original discovery. A total of 34 solar electron events were observed on the Mariner satellites between March, 1964 and August, 1966. These varied in flux from 10 to 6000 electrons/cm²-sec-ster with energies greater than 40 keV. They were associated with small and large flares alike.

These electrons are probably accelerated in the vicinity of the sun and produce the X-rays observed in solar flares by bremsstrahlung in the sun's atmosphere. They are probably responsible for the radio wave bursts that occur at the same time as the X-rays. Representative X-ray fluxes in typical flares are 2×10^{-6} erg/cm²-sec for times of 100 sec. This gives a time integrated X-ray emission from the sun of 5×10^{23} erg. In the sun's atmosphere of 90 per cent hydrogen and 10 per cent He, bremsstrahlung should occur with an efficiency of about 10^{-4}. If 40 keV electrons were responsible for the bremsstrahlung, the total electron source would be 10^{35} electrons. On May 25, 1965 it was estimated[48] that 10^{34} electrons were emitted from the sun. It would thus appear that about 10 per cent of the electrons responsible for the X-rays escape from the sun.

At least 50 per cent more electrons come from the sun's direction than from the opposite direction.[49] This was interpreted as verification that the electrons followed the Archimedes Spiral magnetic field lines through interplanetary space from the sun to the earth.[36] And then arrived at the earth in a tight column about the magnetic field direction.

4.11.7 Neutrons

Severny[51] suggested a model for acceleration of solar flare protons that starts at the neutral point formed in the lower corona between two antiparallel magnetic fields. He considers the instability caused by a change in the position of the sunspots or their field strengths. The changing magnetic field acts like a piston compressing the plasma around a neutral point. The shock front moves inside the contracting plasma. Behind the front a hot dense plasma with a temperature of 3×10^7 °K and a density of 10^{14} particles/cm³ is formed. The plasma is heated by transforming the magnetic sunspot energy first into kinetic energy of accelerated mass motion and then into heat energy.[52] The mass of plasma with its own magnetic field is pushed very rapidly between the converging shocks. The reflected shock stops the contracting magnetic walls and contraction is replaced by an expansion that gives rise to an outgoing blast-wave. In the hot region behind the reflected shock-wave, thermonuclear reactions can occur. With a deuterium abundance[53] of $D/H = 5 \times 10^{-5}$, as many as 10^3–10^4 (d, d) reactions/cm³ could be produced during the flare. The charged particle products of these reactions are then accelerated[52] to higher energies, perhaps as high as 10 BeV, by reflections from the converging magnetic field walls. These are the solar flare protons. If his ideas are correct there will be an observable number of 2.5 MeV neutrons associated with the (d, d) reaction.

$$d + d \rightarrow He^3 + n \qquad (4.19)$$

There may also be 14 MeV neutrons associated with the (d, t) reaction.

$$d + t \rightarrow He^4 + n \qquad (4.20)$$

The tritium is produced with a proton in the (d, d) companion reaction to equation 4.19.

$$d + d \rightarrow t + p \qquad (4.21)$$

Even if thermonuclear reactions do not occur in the corona, neutrons must be made from the bombardment of the sun's atmosphere by solar flare protons.[54] In the acceleration process the protons can make neutrons by the bombardment of He^4 that makes up 10 per cent of the sun's atmosphere. This

$$p + He^4 \rightarrow p + n + He^3 \qquad (4.22)$$

reaction has a threshold of 25.7 MeV. Lingenfelter and associates[55] calculated the flux of neutrons expected near the earth from this model. They included the loss from decays in flight. The resulting number of neutrons per cm² per MeV per incident proton/cm² is given in Fig. 4.30. Three curves are drawn for the different rigidity constants R_0 in the exponential proton spectra of equation 4.18. The three curves peak between 30 and 80 MeV. If the neutrons are released in a time short compared to the transit time to the earth, the sun–earth distance of 1.5×10^8 km acts like a time of flight spectrometer. The fastest neutrons arrive early, the slower ones late. The protons take circuitous paths following the spiral magnetic field lines from the sun to the earth, but the neutrons take straight line paths. Therefore, the fast neutrons arrive before the protons. These neutrons may furnish a possible early warning of solar proton flares.

Simpson[56] suggested that the continuous flux of cosmic ray protons below 200 MeV might be due to the decay of neutrons emitted by the

G

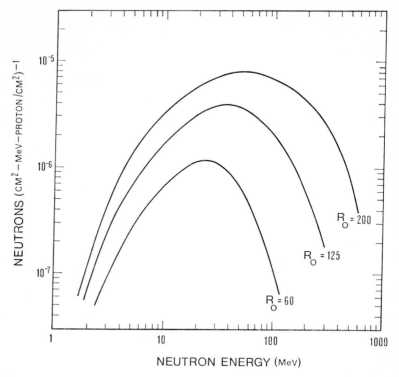

Figure 4.30. Theoretical neutron flux expected near the earth from a large solar flare. Three different source proton rigidities are drawn. The curves bend over at low energies because of the neutron decays in flight from the sun to the earth. They bend over at high energies because of the decrease of source protons at the sun. The curves are normalized to one proton/cm² at the sun. Note that the maximum numbers of neutrons expected at the earth are at energies from 30 to 80 MeV. From Lingenfelter, Flamm, Canfield, and Kellman,[55] 1965.

sun. Dick Lingenfelter and Edith Flamm[57] computed the neutron spectrum at the sun that would account for the observed cosmic ray proton flux. They then transformed the solar neutron spectrum to that expected at the earth. The spectrum peaked at 50 MeV. This flux of neutrons competes favorably with the flux calculated from cosmic ray albedo neutron decay injection (see chapter 1).

High tritium and He³ concentrations were reported in material recovered from Discoverer 17.[58] The gases were attributed to nuclear reactions on the solar surface during the November 12, 1960 flare, perhaps produced by the shock wave thermonuclear heating.

Neutrons from the sun have not yet been identified with certainty.

A balloon flight[59] in 1964 found a day-to-night ratio of 1.00 ± 0.01 at 41 deg N latitude. This gave an upper limit of 0.02 solar neutrons/cm²-sec from 1 to 14 MeV. A ratio of 1.000 ± 0.001 for $1 < L < 1.5$ and 1.050 ± 0.008 for $1.5 < L < 2.2$ was found with a pair of moderated BF_3 proportional counters on OSO–1.[60] The later result is probably caused by secondary neutrons from cosmic ray protons, as all satellite experiments to date have been bothered by local production of a large number of neutrons in the spacecraft.

Experimenters[61] on the Vela satellites searched for solar flare neutrons with He^3 gas proportional counters surrounded with polyethylene moderators. Neutrons with energies below about 20 MeV could be detected. For these neutrons an upper limit to the quiet time neutron flux of 0.01 neutrons/cm²-sec was found. No solar flare neutron bursts with integrated intensities over the threshold of 1.5×10^3 neutrons were detected.

R. Daniel and his colleagues[62] reported a flux of 0.0148 ± 0.060 neutrons/cm²-sec between 20 and 60 MeV from the sun on a balloon flight 6 to 12 hours after an optical flare of magnitude 3 at 3:00 a.m. local time in India on March 23, 1962. However, this experiment would have been more convincing if, in addition, a lack of solar neutrons were found on a night flight.

The same group[63] also used a spark chamber triggered by scintillators to detect the recoil proton or carbon disintegration fragments from neutron collisions in paraffin. The experiment was guarded on 5 sides by anticoincident scintillators but was open on the 6th side for photography of the spark chamber. The experiment was flown in a balloon on April 15, 1966 from Hyderabad, India. The authors claim to have obtained for the first time positive evidence for the emission of high energy neutrons from the sun. The average flux of solar neutrons of 50 to 500 MeV for the 1.5 hour of observation was 0.1 neutrons/cm²-sec. There was a plausible association between the neutron event and a subsolar flare. As with the emulsion experiment, an additional flight with a negative solar neutron signal at night would be comforting.

E. Roelof[64] used the number of protons in interplanetary space to deduce an upper limit to the number of quiet time solar neutrons. He said, suppose protons of 20 to 75 MeV in interplanetary space observed by the experiment on Imp 1[65] are due to solar neutron decays in flight. What is the contribution to the trapped protons in the earth's radiation belts? He concluded that no significant contribution could come from solar neutrons.   Otherwise the proton interplanetary flux would have been higher.

Stephen Holt[66] pointed out that the same type of arguments also make the experiments,[62,63] that reported measuring solar flare neutrons from the sun, in disagreement with interplanetary space proton measurements.   Under the conditions that give the most planetary protons from solar neutron decays, a flat solar neutron spectrum and no interplanetary magnetic field, the experimental neutron flux should have been only 1/3 that observed.   Under more typical conditions, a steep solar neutron spectrum and an interplanetary magnetic field, the neutron flux should have been only 1/100 of that reported.   We conclude that the neutron results are contradictory and it is questionable whether solar neutrons have ever been detected.   Additional experiments to detect and measure the energies of the neutrons are definitely required.

4.11.8 Solar flare theories

A typical solar flare covers about 10^{19} cm², a region that is about one-thousandth of the sun's total area.   The flare occurs over a vertical height of about 10^4 km so that the total volume of the flare is about 10^{28} cm³.   In a large solar flare more than 10^9 protons/cm² with energies greater than 100 MeV arrive at the earth.   This energy at the solar flare source amounts to 10^{32} ergs.   Dividing by the volume we obtain the energy density in the flare region of 10^4 erg/cm³.

What is the source of this energy in the flare region?   One immediately thinks of the energy in thermal motion in the upper chromosphere or lower corona.   The thermal energy density is

$$E = nkT \qquad (4.23)$$

where n is the number of atoms/cm³, k is Boltzman's constant 1.6×10^{-16} erg/atom-°K, and T is the temperature in deg Kelvin. Using the values for n and T derived from Table 1, the energy density at 20,000 km height in the upper chromosphere is only 0.2 erg/cm³ and in the corona at 1 million km height is even less, 10^{-3} erg/cm³.   Consequently, even if all of the thermal energy in the flare volume is transformed into flare energy, the energy density is too low by a factor of 10^4 in the chromosphere and 10^7 in the corona.

The only other source of energy readily available to the flare is the

magnetic field energy.   If a magnetic field of 500 gauss can be transformed into flare energy, the energy released is

$$E = B^2/8\pi \tag{4.24}$$

or 10^4 ergs/cm^3.   The magnetic field energy is sufficient to supply the
solar flare energy.   The problem now shifts to that of finding a mechanism to transform magnetic field energy into particle kinetic energy.

Sweet[67,68] showed that neutral points, lines or planes must occur
above complex sunspots as shown in Fig. 4.31.   The magnetic field
energy density is low or zero near the neutral plane where the magnetic
field is low or zero.   The magnetic pressure is greater away from the
neutral plane.   That magnetic pressure causes plasma with the attached
magnetic field lines to flow toward the neutral plane as indicated in
Fig. 4.32.[68]   The magnetic field lines on either side of the neutral plane
are oppositely directed.   As the plasma approaches the neutral plane,
the magnetic field lines are annihilated, energy is released and the gas
squirts out the ends as indicated.   The magnetic field energy is turned
into particle kinetic energy.   There is one disturbing problem.   Parker[69]
pointed out that the process takes too much time, much longer than the
rise time of a solar flare of 10^2 to 10^3 seconds.   Under the assumption

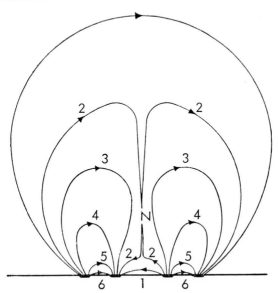

Figure 4.31. Sweet's mechanism for converting magnetic field to particle energy.
Lines from two aligned, bipolar spot pairs annihilate at the neutral point N.
From Sweet,[68] 1958.

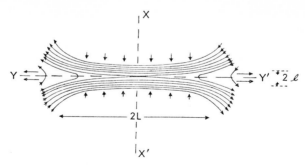

Figure 4.32. Sweet's mechanism for compressing solar plasma. The plasma approaches the neutral plane YY' along the direction XX' as indicated by the arrows. The magnetic field annihilates along YY' and the reconnected magnetic field lines and compressed plasma escape in the direction Y and Y'. $2L$ is the width of the incoming gas and $2l'$ is the width of the escaping plasma. From Parker,[69] 1963.

of stationary flow where the thermal pressure in the neutral plane balances the external magnetic pressure the calculated dissipation time is about 50 hours, more than two orders of magnitude too long.

A similar scheme for the annihilation of magnetic field lines was suggested by Gold and Hoyle.[70] Tubes of twisted magnetic field lines migrate across the surface of the sun and occasionally annihilate tubes having oppositely directed fields. The argument proceeds as with the Sweet Annihilation.

Petschek[71] in 1964 saved the day when he pointed out that previous analyses had overlooked an important mechanism for transforming magnetic field energy into particle kinetic energy. Standing magneto-hydrodynamic waves convert the energy rapidly. At large distances they are responsible for dissipation of the magnetic field energy. At small distances the previously discussed diffusion is the major mechanism. Combining the two mechanisms he found that the energy could be converted in 10^2 seconds, a time in agreement with the observed rise times of solar flares. The dissipation time was no longer a legitimate objection to the Sweet magnetic field annihilation mechanism as the source of the solar flare energy.

The Severny solar flare theory discussed in section 4.11.7 is an example of a non-stationary state approach. In his theory, shock waves in the vincinity of the neutral plane heat up the plasma to a density of 10^{14} atoms/cm^3 at a temperature of 3×10^7 °K, which is sufficient for thermonuclear reactions to take place.

Because the conductivity of the sun's plasma is very high, many theories take the limiting case of the conductivity equal to infinity. In that case an electric field cannot be set up or maintained. However, Alfvén and Carlqvist[72] have proposed an electrical source of solar flare energy in analogy with an inductive electric circuit where there is a general tendency for the energy to concentrate at the point where the circuit is broken. Large energy releases are familiar to the high-power transmission engineers where switches are designed to take most of the magnetic energy of the whole network. Or to the physicists who design superconducting coils for strong magnetic fields where the whole magnetic energy of the circuit is released as an explosion at the point where the conductor ceases to be superconducting.

In the solar atmosphere, such an explosion is expected when the maximum plasma current is exceeded. In the sun's high atmosphere currents follow magnetic field lines. Local constrictions due to the pinch effect could increase the current density in small regions. If the current density exceeds the maximum permitted by the plasma, a local decrease in plasma density results. A high voltage is produced over the high vacuum region and a major fraction of the total magnetic energy in the current circuit is then released near the instability. This is a solar flare. Possible values for the quantities are: a current I of 10^{11} amp, an inductance L of 10 Henries, a time constant τ of 10^3 sec and a voltage drop over the interruption

$$V = L\frac{dI}{dt} \simeq \frac{LI}{\tau} \tag{4.25}$$

of 10^9 volts. The total magnetic energy released by the circuit is

$$E_m = \tfrac{1}{2}LI^2 \tag{4.26}$$

which amounts to 10^{23} Joules or about 10^{30} ergs. The energy release approaches that required for solar flares. In addition, it has the attractive feature that the potential of 1 BeV can accelerate the solar protons to the high energies observed at the earth.

An additional mechanism for accelerating electrons or protons up to high energies is the Fermi Acceleration. Just as a particle with momentum less than 14 BeV/c is turned around at the earth's equator by the earth's magnetic field, a particle in space is reflected by a sufficiently strong magnetic field. If a magnetic wall is moving with a velocity v, then the particle is speeded up or slowed down by $\pm v$, depending on whether the magnetic wall is approaching or retreating. A particle between two approaching magnetic walls will be continuously accelerated as it reflects back and forth between the walls. As the particle picks up energy its longitudinal velocity increases and its pitch angle decreases. Unless an additional process like Coulomb scattering occurs, the particle will eventually penetrate the mirror and escape the acceleration region. The Fermi Acceleration may be the process that accelerates solar protons up to energies of 1 BeV.

To date the experimental data that clarify the energy transformations and particle accelerations in solar flares are so meager that much of the solar flare theory is little more than speculation. The major theories have been discussed in a Review article by Wentzel.[73] In the next few years, particularly over the next solar cycle with its maximum in 1970, the experimental data obtained on satellites above the earth's atmosphere and outside the earth's magnetic field should lead to an understanding of the mysteries of the energy sources, accelerations and interactions in solar flares. We can look forward to a period of thrilling discoveries in solar physics that will surpass even our recent exciting discoveries of the earth's environment.

REFERENCES

1. Unsöld, A., "Solar physics," from *Space Age Astronomy*, Academic Press, Inc., New York, New York, 161–170 (1962).
2. Goldberg, L., Muller, E. A. and Aller, L. H., University of Chicago Press (1960). Data from Goldberg, L., Muller, E. A. and Aller, L. H., "The abundances of the elements in the solar atmosphere," *Astrophys. J. Suppl.* **5**, 1–137 (1960).
3. Burbidge, E. M., Burbidge, G. R., Fowler, W. A. and Hoyle, F., "Synthesis of the elements in stars," *Reviews of Modern Physics* **29**, 547–650 (1957).
4. Brandt, J. C. and Hodge, P., *Solar System Astrophysics*, McGraw–Hill Book Co., Inc., New York, New York (1964).
5. Bahng, J. and Schwarzschild, M., *Astrophys. J.* **134**, 312 (1961).
6. Leighton, R. B., "The solar granulation," from *Annual Review of Astronomy and Astrophysics*, Vol. 1, 19–40 (1963). Published by Annual Reviews, Inc., Palo Alto, California.
7. Leighton, R. B., Noyes, R. W. and Simon, G. W., *Astrophys. J.* **135**, 474 (1962).
8. Neupert, W. M., Gates, W., Swartz, M. and Young, R., "Observation of the solar flare X-ray emission line spectrum of iron from 1.3 to 30 A," *Astrophys. J.* **149**, 179 (1967).
9. Babcock, H. W., "The sun's magnetic field," from *Annual Review of Astronomy and Astrophysics* **1**, 41–58 (1963). Published by Annual Reviews, Inc., Palo Alto, California.

10. Zhulin, I. A., Ioshpa, B. A. and Mogilevskiy, E. I., "Solar magnetic fields," *Geomagnetism and Aeronomy* **2**, No. 4, 489–520 (1962).
11. Hale, G. E., *Astrophys. J.* **28**, 315 (1908).
12. Hale, G. E. and Nicholson, S. B., "Magnetic observations of sunspots," Carnegie Institute of Washington Publ., No. 498, Parts I and II, Washington, D.C. (1938).
13. Babcock, H. W., *Astrophys. J.* **118**, 387 (1953).
14. Babcock, H. W. and Babcock, H. D., *Astrophys. J.* **121**, 349 (1955).
15. Maunder, E. W., *Monthly Notices, Roy. Astron. Soc.* **82**, 534 (1922).
16. Babcock, H. W., *Astrophys. J.* **133**, 572 (1961).
17. Mayfield, Earle, Private communication, Solar Observatory, El Segundo, California (1967).
18. Waldmeier, M., *The sunspot activity in the years 1610-1960*, Schulthess and Co., A.G., Zurich (1961).
19. Mayfield, Earle, Private communication, Solar Observatory, El Segundo, California (1967).
20. Hinteregger, H. E., GRD Rocket Monochrometer, 23, August, 1961, White Sands. (See reference 4, p. 182).
21. Rugge, H. R. and Walker, A. B. C., Jr., "Solar X-ray spectrum below 25 A," *Space Research* **8**, Mitra, A. P. (ed.), North-Holland Publishing Company, Amsterdam, (1968).
22. Chodil, G., Jopson, R. C., Mark, Hans, Seward, F. D. and Swift, C. D., "X-ray spectra from Scorpius (SCO-XR-1) and the sun observed above the atmosphere," *Phys. Rev. Letters* **15**, 605–608 (1965).
23. Kraus, John D., *Radio Astronomy*, McGraw–Hill Book Co., New York, N.Y. (1966).
24. Smerd, S. F., "Radio frequency radiation from the quiet sun," *Australian J. Sci. Res.*, **3A**, 34–39 (1950).
25. Hundhausen, A. J., Asbridge, J. R., Bame, S. J. and Strong, I. B., "Vela satellite observations of solar wind ions," *J. Geophys. Res.* **72**, 1979–1993 (1967).
26. Kreplin, R. W., Chubb, T. A. and Friedman, H., "X-ray and Lyman-Alpha emission from the sun as measured from the NRL SR-1 satellite," *J. Geophys. Res.* **67**, 2231–2253 (1962).
27. Arnoldy, R. L., Kane, S. R. and Winckler, J. R., "Energetic solar flare X-rays observed by satellite and their correlation with solar radio and energetic particle emission," *Astrophys. J.*, 151, 711–736 (1968).
28. Chubb, T. A., Kreplin, R. W. and Friedman, H., "Observations of hard X-ray emission from solar flares," *J. Geophys. Res.* **71**, 3611–3622 (1966).
29. Hey, J. S., *Nature* **157**, 47 (1946).
30. Wild, J. P., Smerd, S. F. and Weiss, A. A., "Solar bursts," from *Annual Review of Astronomy and Astrophysics* **1**, 291–366 (1963). Annual Reviews, Inc., Palo Alto, California.
31. Ellison, M. A. and Conway, M., *Observatory* **70**, 77 (1950).
32. Webber, W. R., "The variations of low energy cosmic rays during the recent sunspot cycle," *Progress in Elementary Particle and Cosmic Ray Physics* **6**, 75–243 (1962) North-Holland Publishing Co., Amsterdam.
33. Hoffman, D. J. and Winckler, J. R., "The time variations of solar cosmic rays during the September 3, 1960, event," *J. Geophys. Res.* **68**, 2067–2098 (1963).
34. Bryant, D. A., Cline, T. L., Desai, U. D. and McDonald, F. B., "Explorer 12 observations of solar cosmic rays and energetic storm particles after the solar flare of September 28, 1961," *J. Geophys. Res.* **67**, 4983–5000 (1962).

35. Webber, W. R., "A review of solar cosmic ray events," in *AAS-NASA Symposium on the Physics of Solar Flares*, NASA SP-50, Technical Information Division, National Aeronautics and Space Administration, Washington, D.C. (1964).

36. Parker, E. N., *Interplanetary Processes*, p. 138, Interscience Publishers, Inc., New York (1963).

37. McCracken, K. G., "Deductions regarding the interplanetary magnetic field," *J. Geophys. Res.* **67**, 447–458 (1962).

38. Webber, W. R., "An evaluation of the radiation hazard due to solar particle events," unpublished (1963).

39. Freier, P. S. and Webber, W. R., "Exponential rigidity spectrums for solar flare cosmic rays," *J. Geophys. Res.* **68**, 1605–1629 (1963).

40. Biswas, S., Fichtel, C. E., Guss, D. E. and Waddington, C. J., "Hydrogen, helium, and heavy nuclei from the solar event on November 15, 1960," *J. Geophys. Res.* **68**, 3109–3122 (1963).

41. Biswas, S., Fichtel, C. E. and Guss, D. E., "Solar cosmic-ray multiply charged nuclei and the July 18, 1961 solar event,"*J.Geophys. Res.***71**, 4071–4077 (1966).

42. Bartels, J., *Terrest. Magnetism Atm. Elec.* **37**, 48 (1932).

43. Bryant, D. A., Cline, T. L., Desai, U. D. and McDonald, F. B., "New evidence for long-lived solar streams in interplanetary space," *Phys. Rev. Letters* **11**, 144–146 (1963) and "Continual acceleration of solar protons in the MeV range," *Phys. Rev. Letters*, **14**, 481–484 (1965).

44. Fan, C. Y., Gloeckler, G. and Simpson, J. A., "Protons and helium nuclei within interplanetary magnetic regions that co-rotate with the sun," *Session Accel.* **3**, *Proc. Intern. Conf. Cosmic Rays, 9th*, London, 1965. The Physical Society, London (1966).

45. Lin, R. P., Kahler, S. W. and Roelof, E. C., "Solar flare injection and propagation of low energy protons and electrons in the event of 7–9 July, 1966," *Solar Physics* **4**, 338–360 (1968).

46. Arnoldy, R. L., Hoffman, R. A. and Winckler, J. R., "Solar cosmic rays and soft radiation observed at 5,000,000 kilometers from earth," *J. Geophys. Res.* **65**, 3004–3007 (1960).

47. Hoffman, R. A., Davis, L. R. and Williamson, J. M., "Protons of 0.1 to 5 MeV and electrons of 20 keV at 12 earth radii during sudden commencement on September 30, 1961" *J. Geophys. Res.* **67**, 5001–5006 (1962).

48. Van Allen, J. A. and Krimigis, S. M., "Impulsive emission of 40-keV electrons from the sun," *J. of Geophys. Res.* **70**, 5737–5751 (1965).

49 Anderson, K. A. and Lin, R. P., "Observations on the propagation of solar-flare electrons in interplanetary space," *Phys. Rev. Letters* **16**, 1121–1124 (1966).

50. Lin, R. P. and Anderson, K. A., "Electrons > 40 keV and protons > 500 keV of solar origin," *Solar Physics* **1**, 446–464 (1967).

51. Severny, A. B., "Solar flares," *Annual Review of Astronomy and Astrophysics*, Vol. 2, p. 363–400, Annual Reviews, Inc., Palo Alto, Calif., (1964).

52. Severny, A. and Shabansky, V., *Izv. Krymsk. Astrofiz. Obs.*, **25**, 88 (1961).

53. Severny, A., *Izv. Krymsk. Astrofiz. Obs.* **16**, 3 (1956).

54. Hess, W. N., "Neutrons in space," *Proceedings of the Fifth Inter-American Seminar on Cosmic Rays*, La Paz, Bolivia, 17–27, July (1962).

55. Lingenfelter, R. E., Flamm, E. J., Canfield, E. H. and Kellman, S., "High energy solar neutrons I. production in flares and II. flux at the earth," *J. Geophys. Res.* **70**, 4077–4095 (1965).

56. Simpson, J. A., in Semaine d'étude sur le problème du rayonnement cosmique dans l'espace interplanétaire (Pontificia Academia Scientiaryum, Vatican, 1963) p. 323–352.

57. Lingenfelter, R. E. and Flamm, E. J., "Solar neutrons and the earth's radiation belts," *Science* **144**, 292–294 (1964).

58. Fireman, E. L., "Solar surface nuclear reactions," in *AAS-NASA Symposium on the Physics of Solar Flares*, NASA SP-50, Scientific and Technical Information Division, National Aeronautics and Space Administration, Washington D.C., 279–284 (1964).

59. Haymes, R. C., "Terrestrial and solar neutrons," *Rev. Geophys.* **3**, 345–364 (1965).

60. Hess, W. N. and Kaifer, R. C., "Search for solar neutrons," *Trans. Am. Geophys. Union* **44**, 83 (1963).

61. Bame, S. J. and Asbridge, J. R., "A search for solar neutrons near solar minimum," *J. Geophys. Res.* **71**, 4605–4616 (1966).

62. Apparao, M. V. K., Daniel, R. R., Vijayalakshiri, B. and Bhatt, V. L., "Evidence for the possible emission of high energy neutrons from the sun," *J. Geophys. Res.* **71**, 1781–1785 (1966).

63. Daniel, R. R., Joseph, G., Lavakare, P. J. and Sunderrajan, R., "High energy neutrons from the sun," *Nature* **213**, Jan. 7, 21–23 (1967).

64. Roelof, E. C., "Effect of the interplanetary magnetic field on solar neutron-decay protons," *J. Geophys. Res.* **71**, 1305–1317 (1966).

65. McDonald, F. B. and Ludwig, G. H., "Measurement of low-energy primary cosmic ray protons on Imp 1 satellite," *Phys. Rev. Letters* **13**, 783–785 (1964).

66. Holt, S. S., "Satellite protons as a test of solar neutron observations," *J. Geophys. Res.* 3507–3510 (1967).

67. Sweet, P. A., "The production of high energy particles in solar flares," *Nuovo Cimento* **8**, 188–196 (1958).

68. Sweet, P. A., *Electromagnetic phenomena in cosmical physics*, Lehnert B. (ed.), Cambridge Univ. Press, 123–134 (1958).

69. Parker, E. N., *Astrophys. J. Suppl.* **8**, 177–212 (1963).

70. Gold, T. and Hoyle, F., *Mon. Not. Roy. Astr. Soc.* **120**, 89–105 (1960).

71. Petschek, Harry E., "Magnetic field annihilation," in *AAS-NASA Symposium on the Physics of Solar Flares*," NASA SP-50, Scientific and Technical Information Division, National Aeronautics and Space Administration Washington, D.C., 425–437 (1964).

72. Alfvén, H. and Carlquist, P., "Currents in the solar atmosphere and a theory of solar flares," *Solar Physics* **1**, 220–228 (1967).

73. Wentzel, D. G., "The origin of solar flares and the acceleration of charged particles," in *AAS-NASA Symposium on the Physics of Solar Flares*, NASA SP-50, Scientific and Technical Information Division, National Aeronautics and Space Administration, Washington, D.C., 398–407 (1964).

CHAPTER 5

Interplanetary Space

5.1 INTRODUCTON

Since Gallileo's observations of the planets with the newly invented
telescope in 1609, man has speculated about the properties of interplanetary space. The speculations were little more than guessing until
the discovery of the proton and the electron and the invention of the
cathode ray tube in the 1890's. As early as 1892 the British physicist
George Fitzgerald, better known for the Fitzgerald contraction of
special relativity, discussed the relation between sunspots and magnetic
storms in the *Electrician*, ". . . a sunspot is a source from which some
emanation like a comet's tail is projected from the sun. . . . Is it
possible, then, that matter starting from the sun with the explosive
velocities we know possible there, and subject to an acceleration of
several times solar gravitation, could reach the earth in a couple of
days?"[1]

Eight years later in the same journal he writes: "There are many
things which seem to show that comets' tails, aurorae, and solar
corona, and cathode rays are closely allied phenomena."[2] And he
says about comets, "In connection with the theory of comets' tails I
would call attention to the difficulty of explaining their repulsion by the
sun to the Maxwell pressure of radiation in that there is no evidence
that the molecules of any gas absorb more than a very minute pro-
portion of the radiations that fall upon them."

During this same period of time Kristan Birkeland, a Norwegian
physicist, studied the deflection of a beam in a cathode ray tube with a
magnet.[3] He became convinced that "The luminous pattern in the air
of the tube . . . shows an astonishing analogy with certain cases of
stratification of the aurora borealis at our (Norwegian) latitudes." He
continued that these cathode rays "come from cosmic space and are

absorbed principally at the terrestrial magnetic poles; one must attribute these rays, in one manner or another to the sun. Thus would one explain the daily occurrence of the aurora borealis always in the arctic countries, and also the well known coincidence between the eleven-year period of the solar activity and the aurora borealis."

Kristan Birkeland continued his interest in the aurora both by performing experiments in the laboratory on his model terrella (little earth) and with expeditions into the arctic. In two classic volumes he describes the results of the 1902–1903 Norwegian Auroral Expedition.[4] He organized the largest geomagnetic data analysis program up to that time and obtained magnetograms and auroral observations from 25 observatories throughout the world. He inspired the auroral and cosmic-ray work of his colleague Carl Stoermer who spent more than 40 years in the study of the trajectories of charged particles in the earth's magnetic field. Alex Dessler,[5] describes the adventurous spirit of the geophysicists of those days in an interesting review article of the solar wind and interplanetary magnetic field.

During the golden era of atomic and nuclear physics that followed from 1900 to the 1940's, progress in the understanding of interplanetary space was slow. Data on magnetic disturbances on the earth and ionospheric disturbances for radio waves was steadily accumulated. Outstanding among the few dedicated geophysicsts who continued to contribute heavily during this time was Sydney Chapman. This year (1968) he is celebrating his 82nd birthday and is still active as he commutes between the University of Alaska and the High Altitude Observatory at Boulder, Colorado. He teamed with the German geophysicist Julius Bartels, the world's expert on magnetic storms, to write their monumental work[6] on geomagnetism in 1940.

Sydney Chapman and V. C. A. Ferraro[7] in 1934 were firmly convinced that charged particles were emitted from the sun at times of magnetic storms and that the interaction of these particles with the earth's magnetic field formed a magnetic cavity around the earth and a magnetic tail on the dark side of the earth away from the sun. They calculated that the ions traveled at 1000 km/sec and reached the earth in 2 days. These charged particles streaming toward the earth caused the changes in the earth's magnetic field.

Studies by Scott Forbush[8] in 1938 showed that the number of cosmic ray protons striking the earth varied over the solar cycle. The number of cosmic rays was greatest during the minimum of the solar cycle and smallest during the maximum. Decreases in the number of cosmic ray protons were also observed during the magnetic storms. These decreases are now called Forbush decreases after their discoverer. When the solar plasma was thought to be most intense the number of cosmic ray protons was the least. Philip Morrison[9] of Cornell University furnished the explanation. He suggested that the solar plasma carried along some of the sun's magnetic field that swept out the cosmic ray particles in interplanetary space. To reach the earth the cosmic rays had to diffuse through this magnetic field. The repulsion of the cosmic rays was greatest when the solar plasma and its trapped magnetic field was most intense.

Zodiacal light observed with astronomical telescopes probably gave the first indication of material in interplanetary space. Cassini in 1672 suggested that the zodiacal light was caused by the scattering of sunlight from dust particles in the ecliptic plane. However, Behr and Seidentopf[10] measured the polarization of the scattered zodiacal light in the plane of ecliptic and found it to be large—23 per cent. They suggested that the scattering must be caused by electrons as well as dust and deduced a maximum density of 600 electrons/cm^3. Since the zodiacal light seems to come from the plane of the ecliptic at all positions of the earth's orbit,[11] it appeared likely that the zodical light was scattered from the solar corona. In his review article on interplanetary space in 1963, Reimar Lüst[12] concludes that the values of the electron density obtained from zodiacal light measurements are highly uncertain.

The occultation (concealment) of radio sources by the interplanetary electrons was suggested in 1951. Radio waves are scattered by the random fluctuations in the electron densities and are refracted by the electron density gradients. The first effect increases the apparent size of the source and the second changes its position. The occultation has demonstrated the presence of interplanetary electrons but has not given quantitative values for their densities.[12]

It was Ludwig Biermann[13] in 1951 who observed that a stream of charged particles continuously leaves the sun. It flows during magnetic storms and quiet times alike but is stronger during the storms. These conclusions were obtained from studies of comet tails that always point away from the sun. He showed that the light pressure from the sun was not sufficient to cause the tail deflection but that a plasma of electrons and protons could interact with the material of the comet and deflect the tail. His work stimulated additional important theoretical and experimental studies that eventually led to the discovery of the solar emission of electrons and protons. As sometimes happens in science, the prediction of particles from the sun was correct even though it now appears that the deflection of comet tails is caused by a different mechanism—a plasma-magnetic field interaction.[14,15]

Measurements of the temperature of the earth's high atmosphere with rockets in the 1950's showed that the atmosphere became hotter rather than cooler at greater distances from the earth. Sydney Chapman[16] suggested that the heating is caused by conduction of heat away from the sun through the solar corona. He calculated that the heat flow through the ionized hydrogen of the corona should vary with the temperature as $T^{5/2}$ and that the temperature of the corona should fall off only slowly with distance from the sun, $T \propto 1/r^{2/7}$. The temperature in the corona in the vicinity of the earth should be about 2×10^5 °K. He then calculated the density of atoms in the corona using the barometric law, the atmospheric pressure is equal to the weight of the atmosphere above, as in the earth's atmosphere and found a density of 10^2 atoms/cm³. Chapman concluded that the solar corona extended all the way to the earth. It extends outward through interplanetary space and fills the whole solar system.

There now existed two pictures of interplanetary space. Biermann's particles racing away from the sun at high speeds and Chapman's static solar atmosphere that extended beyond the earth. Eugene Parker from the University of Chicago solved the dilemma. In his words, "Now there were apparently two bodies of solar vapor to think about: the steady corona and the stream of particles flowing out from the sun at high speed. This, however, was impossible. In a magnetic field one stream of charged particles cannot pass freely through another, and it was known that solar-system space was filled with magnetic fields. Therefore the corona and the solar stream could not be separate entities. They must be one and the same. The corona, behaving like a static atmosphere near the sun, must become a high-velocity stream farther out in space."[17] Our understanding of interplanetary space today is based upon Eugene Parker's theory of the expansion of the corona.[18] This expansion he named the "Solar Wind."

5.2 THE SOLAR WIND

5.2.1 Theory

Parker's picture starts with an almost stationary proton–electron gas at the bottom of the corona. It moves away from the sun at a few hundred km/sec and is replaced by more gas rising from the photosphere. The gas is slowly accelerated as it moves farther away from the sun. It takes a distance of about 10 sun's radii to reach the speed of a few hundred km/sec and gradually levels off to a rather constant velocity at the earth's orbit. In contrast to the static equilibrium of planetary atmospheres, the sun's atmosphere with the extended temperature in interplanetary space has supersonic expansion, it moves faster than the velocity of sound.

Parker first showed that the usual equations for static equilibrium, the barometric law and the conductivity heat flow equations, with reasonable values of the constants, gave pressures at infinity that were too high.[18] He concluded that it was not possible for the solar corona or the atmosphere of any star to be in hydrostatic equilibrium out to large distances. He then calculated the steady expansion that is expected for the solar corona. His resulting expansion velocity is plotted in Fig. 5.1a against the distance from the sun for different values of the corona temperature. These curves show that the solar wind has nearly reached its plateau velocity by the time it arrives at the earth. At the earth's orbit of 1.5×10^8 km and at a corona temperature of 10^6 °K the solar wind velocity is about 400 km/sec. The calculated solar wind density is shown in Fig. 5.1b. At the earth the density has dropped to about 5 protons/cm³. The subject of the expansion of the corona is extensively treated in the excellent review paper by Eugene Parker.[19] Rather than continue with his detailed mathematical treatment here, we will follow the more descriptive treatment of Clauser,[20] following Alex Dessler,[5] who uses the analogy to the deLaval nozzle.

In a compressible gas flowing through a tube of decreasing cross-section, Fig. 5.2a, the mass flow at equilibrium is a constant.

$$A\rho v \ (\text{gm/cm}^2) = C \tag{5.1}$$

A is the cross-sectional area of the tube, ρ is the density of the gas, v is the velocity of the gas and C is a constant. The flow velocity is limited to the velocity of sound so that further constriction of the tube only decreases the mass flow. If, however, a diverging section is added as in

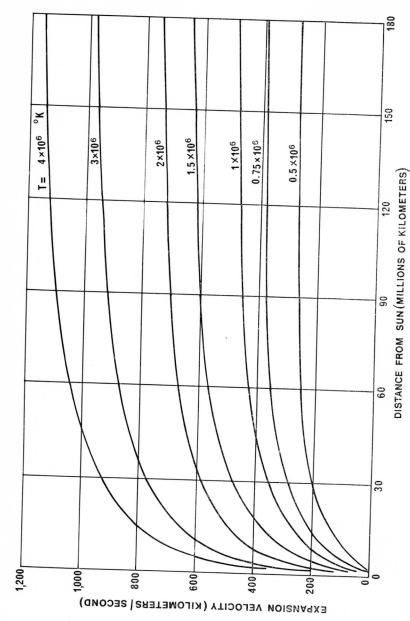

Figure 5.1. a. A plot of the solar wind velocity v_s versus the distance from the sun. Each curve is plotted for a different value of the temperature of the corona. The highest temperatures give the highest expansion velocities. The arrow points to the distance to the earth's orbit. From Parker,[17] 1964.

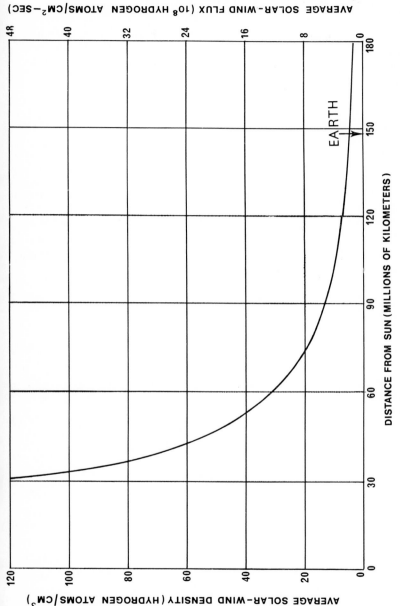

Figure 5.1. b. A plot of the solar wind density and flux versus the distance from the sun. The solar wind density scale is on the left side of the graph and the solar wind flux is on the right. The arrow points to the distance to the earth's orbit. From Parker,[17] 1964.

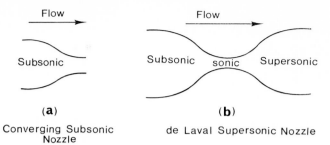

(a)   (b)

Converging Subsonic   de Laval Supersonic Nozzle
      Nozzle

Figure 5.2. a. Converging subsonic nozzle. In such a nozzle the gas velocity can approach sonic velocity in the neck but can never become supersonic.

b. deLaval supersonic nozzle. With the addition of an expanding section beyond the neck the gas velocity can become supersonic. In the solar corona the sun's gravitational force takes the place of the neck of the nozzle in the solar wind expansion.

Fig. 5.2b, supersonic velocities, velocities greater than the velocity of sound can be obtained. For steady state flow the differential pressure, dp, is balanced by the change in momentum/cm²-sec

$$dp = - \rho v \, dv \tag{5.2}$$

For an adiabatic process

$$\frac{dp}{d\rho} = v_n^2 \tag{5.3}$$

where v_n is the velocity of sound. The fractional change in density is

$$\frac{d\rho}{\rho} = - \frac{v}{v_n^2} \, dv \tag{5.4}$$

Substituting Eq. 5.4 into the derivative of the log of Eq. 5.1, we find

$$\left(\frac{v^2}{v_n^2} - 1 \right) = \frac{dA}{A} \bigg/ \frac{dv}{v} . \tag{5.5}$$

For a converging tube dA/A is negative and dv/v is positive so that v must always be less than v_n. For v to exceed the velocity of sound, dA/A must be positive, the tube must increase in area. If the velocity of the gas reaches the velocity of sound in the narrowest part of the tube and it then expands, the nozzle acts as a deLaval nozzle and the velocity exceeds the velocity of sound. If the velocity does not reach the velocity of sound in the most constricted region, the tube acts as a Venturi tube and the gas slows down as it passes out of the nozzle.

If the downstream end of the deLaval nozzle opens into a vacuum, the flow out will always be supersonic.

In the case of the solar corona the sun's gravitational force takes the place of the neck of the deLaval nozzle. It constricts the flow of the solar gas and permits the development of the supersonic flow. Equation 5.2 is replaced by

$$dp = -\rho v \, dv - \rho g \, dr = -\frac{\rho \, M_\odot G \, dr}{r^2}. \tag{5.6}$$

g is the sun's acceleration of gravity. G is the gravitational constant 6.67×10^{-8} dynes-cm²/gm² and $M_\odot$ is the mass of the sun. Taking a radial expansion for the gas, $A \propto 1/r^2$ and $dA/A \propto 2 \, dr/r$, Eq. 5.5 becomes

$$\frac{v^2}{v_n^2} - 1 = \left(2 - \frac{M_\odot G}{v_n^2 r}\right) \frac{(dr/r)}{(dv/v)} \tag{5.7}$$

For dv/v positive and r increasing (dr/r positive), $v < v_n$ if $M_\odot G/v_n^2 r > 2$ and $v > v_n$ if $M_\odot G/v_n^2 r < 2$.

The distance from the sun at which the velocity becomes supersonic is

$$r_n = \frac{M_\odot G}{2v_n^2} \tag{5.8}$$

The supersonic radius may be written in terms of the escape velocity $v_\odot = (2 M_\odot G/r)^{1/2}$ and then

$$r_n = R_\odot \left(\frac{v_\odot}{2v_n}\right)^2 \tag{5.9}$$

where $v_\odot$ is the escape velocity from the surface of the sun. As with Parker's derivation, the force of the magnetic field of the sun and other smaller forces, such as the viscous force, have been neglected. However, the results are in exact agreement with his more elegant mathematical treatment.

The supersonic radius can now be calculated using $v_n = (2\gamma kT/m_p)^{1/2}$. The γ is the ratio of the specific heats at constant pressure to constant volume 5/3, and the 2 comes from the two species of particles, electrons and protons, in thermal equilibrium. The k is the gas constant, 1.38×10^{-16} erg/deg. If we take $T = 10^6$ °K then the supersonic velocity is 1.7×10^7 cm/sec and r_n is $3.5 R_\odot$ If the minimum temperature is greater than 4×10^6 °K, $r_n < R_\odot$ and the expansion is subsonic. However, since the temperature drops rapidly down through the chromosphere to the sun's surface there is always a temperature sufficiently low that $r_n > R_\odot$ and supersonic expansion from the sun always occurs.

An alternative way to see that an atmosphere which is too hot will not expand with supersonic velocity is to look at the density of a static atmosphere

$$\rho(r) = \rho_0 \exp^-\left(\frac{M_\odot m_p G}{kTr}\right) \tag{5.10}$$

When the atmosphere is heated the mass flow outward through a spherical surface at r is

$$\rho(r)v(r)4\pi r^2 = \text{const.} \tag{5.11}$$

If T is very high Eq. 5.11 gives

$$\rho(r) = \rho_0 \left(1 - \frac{M_\odot m_p G}{kTr}\right) \tag{5.12}$$

and $\rho(r)$ is then a constant independent of r. The $v(r) \propto 1/r^2$, a decreasing function of r so $v(r)$ can never become supersonic. However for T much lower, $v(r)$ falls off faster than $1/r^2$ and $v(r)$ must then increase with r and eventually becomes supersonic. The condition, then, for supersonic velocities is that the sun's gravitational force must cause the sun's atmospheric density to decrease faster than $1/r^2$.

The decrease in v with r caused by the sun's gravitational field has not been included in the calculations. The particle's escape velocity from the surface of the sun is 600 km/sec. The maximum exhaust velocity of a mixture of electrons and protons from a deLaval nozzle is $v_{\max} = (2)^{1/2}v_{rms}$. The v_{rms} is the rms speed of the particles in the combustion chamber. For a temperature of 10^6 °K, $v_{rms} = 160$ km/sec. and $v_{\max} = 290$ km/sec. The experimentally observed solar wind velocity is 400 km/sec. The calculated expansion is not even able to furnish the escape velocity let alone the additional observed solar wind velocity.

Saying this another way, the work done to bring one proton–electron pair to the earth's orbit against the earth's gravitational field is 1.9 keV. The kinetic energy of the proton–electron pair with a velocity of 400 km/sec is 0.8 keV. The energy that must be furnished is 2.7 keV/proton–electron pair. The energy available to a proton–electron pair at a temperature of 10^6 °K at the base of the corona is 0.15 keV multiplied by $\gamma = 5/3$ or only 0.25 keV.

The additional energy of 2.7 keV – 0.25 keV = 2.45 keV/proton–electron pair must be furnished by coronal heating. At the present time it is thought that magneto-hydrodynamic waves[21] generated in the photosphere move outward and heat the corona. It is likely that these shock waves furnish the additional heat that is required by the solar wind.

Since the solar wind density falls off as $1/r^2$ from the sun, the solar wind will finally be slowed down and stopped by the magnetic fields and hydrogen of interstellar space. In the direction that the sun is moving the solar wind will feel an opposing pressure caused by the interstellar magnetic field energy density plus the motion of the interstellar streaming hydrogen gas. In the opposite direction the opposing pressure is the interstellar magnetic field energy density only. The latter case gives an interstellar pressure p_i at the distance $r_{\rm sh}$ where the solar wind goes through a shock transition from supersonic to subsonic flow

$$p_i = n_s m_p v_s^2 / (r_{\rm sh}/r_s)^2 \tag{5.13}$$

The n_s is the number of solar wind protons/cm³ at the distance $r_s = 1$ AU, v_s is the solar wind velocity, and m_p is the proton mass. The interstellar magnetic field energy density is

$$p_i = B_i^2 / 8\pi \tag{5.14}$$

where the magnetic field in interstellar space is B_i. We then find that the shock distance is

$$r_{\rm sh}/r_s = v_s \sqrt{8\pi n_s m_p} / B_i \tag{5.15}$$

Substituting $v_s = 4 \times 10^2$ km/sec, $n_s = 5$ protons/cm³, and $B_i = 1\gamma$, we find that $r_{\rm sh} = 60$ AU.

However, in the direction that the sun is moving the additional hydrogen streaming pressure must be added. Then

$$r_{\rm sh}/r_s = \{n_s m_p v_s^2 / [(B_i^2/8\pi) + n_{\rm H} m_{\rm H} v_{\rm H}^2]\}^{1/2} \tag{5.16}$$

Taking $v_{\rm H} = 20$ km/sec, $n_{\rm H} = 1$ hydrogen atom/cm³, and $m_{\rm H}$ the hydrogen atomic mass, the distance to the solar wind shock boundary is $r_{\rm sh} \simeq 30$ AU. It will likely be some time before spacecraft with experiments will pass through the boundary to the solar wind cavity and measure its properties.

5.2.2 Experiments

The launching of deep space probes in 1959 made it possible to measure the densities of ions and electrons in interplanetary space outside the earth's magnetic cavity without the disturbing influences of the earth's magnetic field. The plasma probes detected the ions or electrons on a collector plate after they passed through the two grids. Electrons or positive ions with energies less than a specified amount are repelled by the voltage applied to the first grid. Protons of all energies and electrons with energies greater than 200 eV could be detected. The second grid was held at a negative voltage so that the photoelectrons produced on the collector by the ultraviolet radiation from the sun could be returned to the collector. A background of electrons caused by sunlight reflected from the back of the second grid limited the sensitivity to about 10^8 protons/cm²-sec.

The Soviet satellite, Lunik 2, was launched in December, 1959 in a direction away from the sun into the region of the tail of the earth's magnetic field. At distances from 29 earth radii R_e to impact on the moon at $60R_e$ the plasma probe measured the flux of positive ions with energies greater than 15 eV. The flux was about 2×10^8 particles/cm²-sec.[22]

Lunik 3 was launched in October, 1959 in the general direction of the sun and detected a flux of about 4×10^8 positive ions/cm²-sec at $20R_e$.[23] Venusik was fired in December, 1961, also in the general direction of the sun, and obtained, in at least one case, a flux of 10^9 positive ions/cm²-sec.[23]

In March, 1961, Explorer 10 was launched away from the sun in a direction of about 40 deg from the sun–earth line. It carried the plasma probe (Faraday Cup) of the MIT group consisting of 4 grids and a collector plate inside a cup.[24] Grids one and three were connected to the space-craft so were held at 0 volts relative to the satellite. Grid 4 was held at a constant potential of -130 volts to suppress the photoelectric emission from the collector plate. Grid 2 was connected to a modulator that varied the voltage between two fixed voltages. In this way the backgrounds of electrons from sunlight reflected by the collector onto the rear surface of the fourth grid were eliminated. The Lunik and Venusik plasma probes were not modulated in this way. A sketch of the Explorer 10 plasma detector is shown in Fig. 5.3.

The trajectory of Explorer 10 was almost tangent to the earth's magnetic tail boundary so that the spacecraft dipped into and out of the earth's magnetic field. During one period of time from March 26 to March 27, 1961, the spacecraft appeared to be in interplanetary space. There, the average flux[24] was 2×10^8 protons/cm²-sec, an average

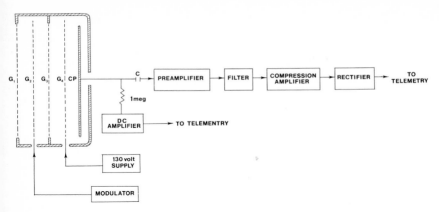

Figure 5.3. Schematic diagram of the MIT plasma probe experiment and electronic system on the spacecraft Explorer 10. From Bridge, Lazarus, Rossi and Sherb,[24] 1963. Grids G_1 and G_3 are at spacecraft ground and G_4 is at -130 V to repel the photoelectrons back to the collector plate CP. G_2 is modulated in phase with the collector to eliminate backgrounds of electrons produced by sunlight reflected by the collector onto the rear surface of G_4.

density of 7 protons/cm³. The solar wind average velocity was 280 km/sec and the corresponding average kinetic energy was 420 eV. The proton temperature obtained from the spread in proton energies was 6×10^5 °K.

The above data tended to verify the existence of the solar wind but was not completely convincing because of problems of detector backgrounds and the locations in interplanetary space where the data were taken. It was only after the Mariner 2 electrostatic analyzer measurements by Conway Snyder and Marcia Neugebauer, from the Jet Propulsion Lab,[25,26] that the evidence was overwhelming and there was universal acceptance of the solar wind. The data were obtained from August 29, 1962 through January 3, 1963. Charged particles with the proper energy per unit charge traveled through the aperture to the collector between two curved plates across which a voltage was applied. The voltage was varied in ten steps so that 10 different energies could be measured. A schematic drawing of the detector is shown in Fig. 5.4.[27]

They found, during quiet times, that the velocity is 460 km/sec, the ion density is 2.5 protons/cm³, and the temperature in the direction of motion is 1.9×10^5 °K. The energy density in particle motion is 4.4×10^{-9} erg/cm³. With an average magnetic field of 5 gamma, the

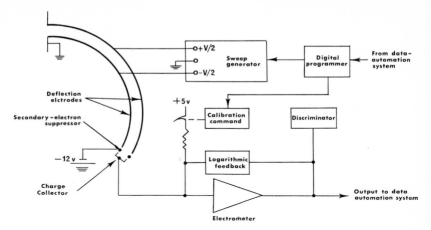

Figure 5.4. Schematic diagram of the Mariner 2 positive-ion spectrometer and electronics. Charged particles with the proper energy/unit charge travel between the curved electrodes to the collector. The voltage across the electrodes was varied in 10 steps so that 10 different energies could be measured. From Neugebauer and Snyder,[26] 1966.

energy in the magnetic field is 1.0×10^{-10} erg/cm³. The ratio of the particle energy density to the magnetic field energy density is 44, so the magnetic field is carried along frozen into the plasma. The Alfvén velocity is 69 km/sec and the ratio of the particle velocity to the Alfvén velocity is 6.7. So the solar wind is clearly supersonic. The results gave complete verification for Parker's theory of the solar wind.

The authors, with U. R. Rao,[27] compared the plasma results with cosmic-ray variations and with solar and geomagnetic activity. They concluded:

"1. There was always a measurably large plasma flow from the direction of the sun.

2. The plasma velocity was not steady. It varied from day to day.

3. There did not seem to be any obvious relationship between plasma velocity and cosmic-ray diurnal amplitude or time of maximum.

4. There was no strong correlation between plasma velocity and overall solar activity as determined by sunspot number and 10.7 cm radiation.

5. There was a very good correlation between K_p and plasma velocity during the entire period of observation which indicates conclusively that high K_p was always related to high velocity plasma.

6. The data fit the equation

$$v \text{ (km/sec)} = 8.44 \sum K_p + 330 \qquad (5.17)$$

where v and $\sum K_p$ are both daily values. . . .

7. The plasma velocity showed a very strong 27 day recurrence tendency and a close association with M region storms, indicating that M regions are emitters of high velocity plasma.

8. No dependence of plasma velocity on solar distance between 1.0 and 0.7 AU could be detected."

With sufficiently good energy resolution, ion detectors observe two peaks in the energy/charge distributions. A plot of such a distribution obtained by the Los Alamos hemispherical plate electrostatic analyzer on Vela 3[28] is shown in Fig. 5.5. The largest peak is made up of the most abundant protons and the smaller peak at exactly twice the energy/charge is made up of He[++] ions. The He[++] ions travel at the same velocity as the protons and electrons in the solar wind. The He[++] peak is very nicely resolved from the proton peak. The ratio of the density of He[++] ions to the density of protons varies with time. The distribution for the time period during August 1965 is given in Fig. 5.6. The ratios vary from 0.01 to 0.15. The average value is 0.042 in good agreement with the value of 0.046 obtained from Mariner 2 during the latter part of 1962.[29] Rarely is the ratio as high as 0.09, the value believed to hold for the solar atmosphere.[30] Much interesting and use-

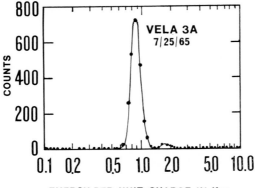

Figure 5.5. The resolution of the Los Alamos hemispherical plate electrostatic analyzer on the spacecraft Vela 3. From Hundhausen, Asbridge, Bame, Gilbert and Strong,[28] 1967. The analyzer clearly separates the singly ionized protons in the large peak from the doubly ionized α-particles in the low peak. The α-particle peak is at exactly twice the energy/unit charge of the proton peak.

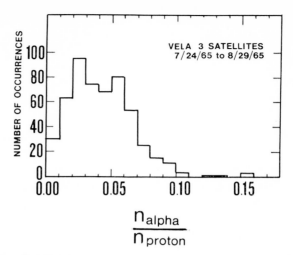

Figure 5.6. The distribution of the measured ratios of α-particles to protons obtained by the electrostatic analyzer on Vela 3. From Hundhausen, Asbridge, Bame, Gilbert and Strong,[28] 1967. The mean value of the ratio is 0.042.

ful data has been obtained from the Vela satellites even though their main mission is to monitor nuclear explosions in space.

Does the mass flow or the energy flow out of the sun in the solar wind contradict our other ideas of energy balance on the sun or the age of the solar system? For quiet times, using a mean flux of 10^8 protons/cm²-sec and energies of 500 eV, the energy in the solar wind at the earth's radius is 2×10^{26} erg/sec. This energy is negligible when compared to the visible light emission of 4×10^{33} erg/sec. The age of the sun is estimated to be about 10^{10} years. In that time 1.4×10^{29} gm has escaped as the solar wind. This is a small fraction, less than 3×10^{-5} of the total mass of the sun.

5.3 MAGNETIC FIELDS

5.3.1 Theory

In the derivation of the properties of the plasma in section 5.2 we ignored the effect of the magnetic fields. This was justified on the grounds that the energy density in the particle motion was large compared to the energy density in the magnetic fields. Observations of the velocity and of the density of the solar wind verified the theory. We are now in a position to ask about the properties of the magnetic fields in interplanetary space.

Eugene Parker[18] predicted that the ionized gas of the solar wind would carry the magnetic field lines outward into interplanetary space. The lines are embedded in the sun at one end and in the gas at the other. If there were no rotation of the sun they would be stretched out radially away from the sun. He writes, "The radial configuration will be as universal as Biermann's outward-gas motion, which is responsible for it . . . Hence, with the more or less steady outflow . . . we expect a radial solar magnetic field falling off approximately as $1/r^2$ in interplanetary space." He then takes the rotation of the sun into account and calculates the expected spiral structure.

In the supersonic plasma of the solar wind a plasma-magnetic signal cannot be transmitted upstream toward the sun. So the supersonic plasma streams away from the sun without exerting a back torque on the sun and the escaping gas does not rotate with the sun.

Sydney Chapman[31] in 1929 first considered the problem of a jet-stream of gas shooting out from the rotating sun. As the sun rotates, the gas leaves radially during each successive time interval. When viewed at a particular time the stream forms a curve in space called an Archimedes spiral. The curve has the same shape as the jet of water thrown from a rotating sprinkler nozzle. The angle between the radius vector and the tangent to the Archimedes spiral is called the "garden hose angle." A snapshot of the Archimedes spiral of the magnetic field line is shown in Fig. 5.7 on the sixth day after the source passed the earth–sun line.

The spiral structure of the magnetic field line in the equatorial plane is described by radial and azimuthal equations

$$r = v_s t + b \tag{5.18}$$

$$\phi = \phi_0 + \Omega_\odot t \tag{5.19}$$

b is the distance from the sun to where the supersonic velocity of the solar wind becomes nearly constant, v_s is the solar wind velocity, $\Omega_\odot$ is the angular velocity of the sun, and ϕ is the sun's longitude measured from the initial longitude ϕ_0. The equation for the spiral motion is

$$\phi - \phi_0 = \frac{\Omega_\odot}{v_s}(r - b) \tag{5.20}$$

obtained by eliminating t and is shown in Figs. 5.7 and 5.8. A finite source width on the sun leads to the finite stream width shown in Fig. 5.7.

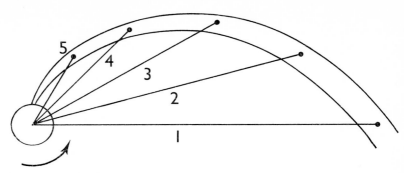

Figure 5.7. The Archimedes spiral magnetic field formed by the solar wind. The straight lines are the trajectories of solar wind particles from the rotating sun in the solar equatorial plane. The diagram is a snapshot after 5 days for particles that started on the sun–earth line on the first day and on the lines labelled by the numbers 2, 3, 4 and 5 on the succeeding days. The curved line is formed by the endpoints of each of the straight lines on the end of the fifth day. Since the sun's magnetic field is carried along and stretched out by the particles, the magnetic field lines have the shape of the Archimedes spiral curves. From Dessler,[5] 1967.

The rotation of the magnetic field lines may be clarified with the analogy to the phonograph needle on a record. The needle moves radially at a constant velocity while the spiral grooves rotate as a rigid body; just as the solar wind moves radially at a constant velocity while the magnetic field rotates as a rigid body.[32]

The direction of the magnetic field expected at the earth is calculated with the help of Fig. 5.9. The earth is located at point E. Angle χ is the angle of a tangent to the magnetic field at the earth with the sun-earth direction. From the diagram we obtain

$$v_s \sin \chi = \Omega_\odot r \cos \chi \tag{5.21}$$

so

$$\tan \chi = \Omega_\odot r / v_s \tag{5.22}$$

Taking $\Omega_\odot$ as 1 rev/27 days (2.7×10^{-6} rad/sec), v_s as 400 km/sec and r as 1.5×10^8 km, we find χ is equal to 45 deg.

Since the magnetic field lines follow the spiral path, the values of the magnetic fields parallel and perpendicular to the stream lines are

$$B_{||} = B_b \left(\frac{b}{r}\right)^2 \tag{5.23}$$

and

$$B_\perp = B_b \frac{\Omega_\odot}{v_s} \left(\frac{b}{r}\right)^2 (r - b) \tag{5.24}$$

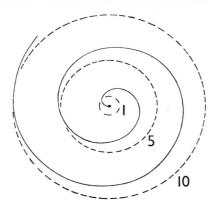

Figure 5.8. The spiral interplanetary magnetic field in the solar equatorial plane. A solar wind velocity of 300 km/sec was used. The dashed circles are drawn at distances of 1, 5 and 10 AU. From Dessler,[5] 1967.

For large r, $r \gg b$, the total magnetic field is

$$B_t = B_b \left(\frac{b}{r}\right)^2 \left(1 + \frac{\Omega_\odot{}^2 r^2}{v_s^2}\right)^{1/2} \tag{5.25}$$

B_b is the magnetic field at the distance b from the sun.

Since the ratio of the particle kinetic energy density to the magnetic field energy density is

$$\beta = \tfrac{1}{2} n_p m_p v_s^2 \bigg/ \frac{B_b^2}{8\pi} \left(\frac{b}{r}\right)^4 \left(1 + \frac{\Omega^2_\odot r^2}{v_s^2}\right) \tag{5.26}$$

and since experimentally $n_p \propto 1/r^2$, we conclude that β is a constant, independent of r, at large r. The particle energy density will then continue to dominate out to large distances from the sun.

5.3.2 Experiments

The magnetic field in interplanetary space was first measured by Paul Coleman, Leverett Davis, and Charles Sonett[33] on the space probe Pioneer V. The magnetic field perpendicular to the spin axis of the spacecraft was obtained with a search coil magnetometer. The spacecraft spin rate was 2.5 rev/sec. The angle between the spin axis of the vehicle and the line to the sun varied from 27 to 2 deg. Over a 50 day time period after the launch on March 11, 1960, magnetic fields from 0.35 to 60 gamma γ were observed. One γ equals 10^{-5} gauss. Whenever conditions were magnetically stable on the earth the average measured

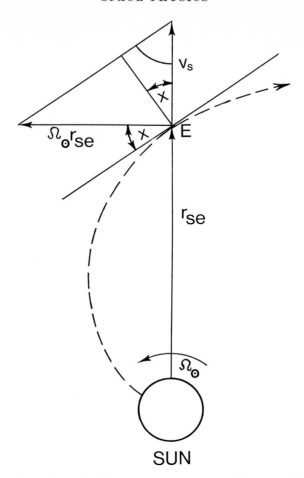

Figure 5.9. Diagram for calculating the angle between the magnetic field line B, and the tangent to the earth at E. The sun rotates with angular velocity $\Omega_{\odot}$. The magnetic field line can be considered as a rigid body rotating with the velocity $\Omega_{\odot}r_{se}$ at the sun–earth distance r_{se}. From Dessler,[5] 1967.

interplanetary magnetic field was near 2.7 γ. The field appeared to be approximately perpendicular to the ecliptic plane. The higher values of 5 to 60 γ were observed when geomagnetic storms occurred on the earth. Especially high values of the interplanetary fields of over 50 gammas occured on April 1 simultaneously with a large magnetic storm on the earth.[34] Unfortunately the magnetometer measured only one component of the magnetic field so that neither the direction nor the magnitude of the total magnetic field in space could be obtained.

Data from later spacecraft show definitely that the magnetic field perpendicular to the ecliptic plane is very small.

The Mariner 2 spacecraft carried a triaxial fluxgate magnetometer with magnetic field sensors in three perpendicular directions.[35]   Although the accuracy of the magnetometer was 1 gamma, the spacecraft residual magnetic field was much larger.   The two residual components perpendicular to the sun–spacecraft direction could be measured and subtracted out.   However, the component of the spacecraft magnetic field parallel to the sun–spacecraft direction still remained.   Measurements were made in late August and early September 1962 and showed that magnetic fields of a few γ are nearly always present.   Magnetic fields, typically about $2\,\gamma$ perpendicular to the sun–spacecraft direction, were present during quiet times and magnetic fields of $5\,\gamma$ during times of small magnetic activity.   The magnetic field rose to $20\,\gamma$ or more during magnetic storms.   Averaged over almost any period of several hours, the measured magnetic field appeared to lie more nearly in rather than perpendicular to the ecliptic plane.

The next attempt to measure the magnetic field in interplanetary space was made with the rubidium vapor magnetometer and two flux gate saturable core magnetometers on Explorer 10.[36]   That satellite was launched in March 1961 in a direction nearly tangent to the boundary of the earth's magnetic tail.   For distances less than $22R_e$ the spacecraft was almost always within the earth's magnetic cavity. For distances from $22R_e$ out to the apogee of $42R_e$ the magnetic field was probably distorted by the interaction between the streaming solar plasma and the magnetic field of the earth's magnetic cavity. Therefore, these measurements were not too useful for obtaining the interplanetary magnetic field.

The first definitive magnetic field measurements in interplanetary space were made by Norman Ness, Clell Scearce and Joseph Seek of NASA-GSFC on Imp 1 (Explorer 18) which was launched on November 27, 1963.   Its initial apogee occurred at $32R_e$ at an angle of 26 deg to the sun–earth direction on the sunlit side of the earth.   The initial satellite spin direction was 111 deg to the sun direction.   Its spin rate was 22 rev/min.   The satellite carried a rubidium 87 vapor magnetometer and two flux-gate magnetometers.   The rubidium vapor magnetometer gave extremely accurate measurements of the interplanetary magnetic field and the two flux-gate magnetometers were used to obtain its direction and magnitude.

H

The rubidium vapor magnetometer uses the Zeeman splitting of the energy ground state of rubidium 87. Optical pumping by a spectral lamp selectively populates one of the Zeeman energy sublevels of Ru^{87} that are separated by the Larmor frequency, 6.99592 cycles/sec-γ, from each other. A weak ac magnetic field at this frequency redistributes the population to all of the levels. The light absorbed in the cell is dependent upon the population of the levels. The resulting modulated light causes the magnetometer to operate as an atomic oscillator at the Larmor resonant frequency. Since the Larmor frequency is proportional to the interplanetary magnetic field, a measurement of that frequency determines the magnetic field. A sketch of the rubidium vapor magnetometer is given in Fig. 5.10.

The flux-gate magnetometer measures the relative but not the absolute magnetic field along its axis. Its saturable magnetic core is driven by a solenoid at 10 kc/sec from positive to negative saturation. The interplanetary magnetic field along its axis as well as the permanent magnetization of the core material generates a second harmonic

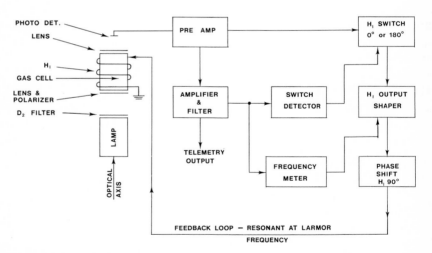

Figure 5.10. Schematic drawing of the self-resonant rubidium vapor magnetometer and electronics for Imp 1. Optical pumping by a spectral lamp selectively populates one of the Zeeman sublevels of Ru^{87} that are separated by the Larmor frequency from each other. A weak ac magnetic field at this frequency redistributes the population to all of the levels. The light absorbed in the cell depends on the population of the levels. The light is modulated and the magnetometer operates as an atomic oscillator at the Larmor resonant frequency. A measurement of the frequency determines the value of the magnetic field. From Ness, Scearce, and Seek,[37] 1964.

signal. When properly calibrated, the voltage caused by the second harmonic signal gives the magnetic field along the flux axis and the phase tells whether the magnetic field is parallel or antiparallel to the axis.

The Imp 1 satellite was especially designed to eliminate spacecraft magnetic fields. They were reduced to less than 1 γ for the rubidium vapor magnetometer and to less than 0.6 γ for the flux-gate magneto-meters. In-flight calibrations indicated that the spacecraft fields were as low as 0.25 γ.

A solar ecliptic coordinate system drawn in Fig. 5.11 was used to present the data. The origin is at the center of the earth, the X_{se} axis points in the direction of the sun, Z_{se} axis is perpendicular to the ecliptic plane and the Y_{se} axis completes the right-handed coordinate system. The magnitude of the magnetic field $\overline{F}$, the latitude θ, and the longitude ϕ are the usual plotted quantities. The variances of the fields from their mean values, δX_{se}, δY_{se} and δZ_{se} are also useful to determine the variability of the magnetic fields.

Magnetic field data are plotted in Figs. 5.12 and 5.13 through four complete days, January 5, 6, 7 and 8. Time is plotted along the bottom of the graphs and the distance from the center of the earth in earth radii

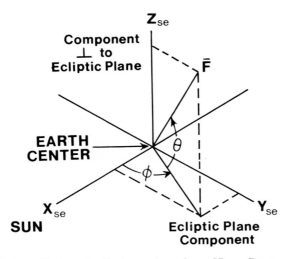

Figure 5.11. Solar ecliptic coordinate system from Ness, Scearce, and Seek,[37] 1964. X_{se} is in the direction from the earth to the sun. Y_{se} is perpendicular to X_{se} and is in the ecliptic plane. Z_{se} is perpendicular to both X_{se} and Y_{se}. $\overline{F}$ is the average value of the magnetic field. θ is the polar angle and ϕ the azimuthal angle of the magnetic field.

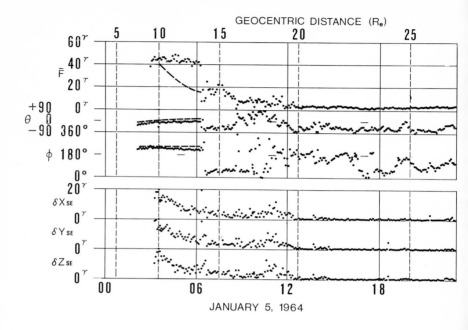

JANUARY 5, 1964

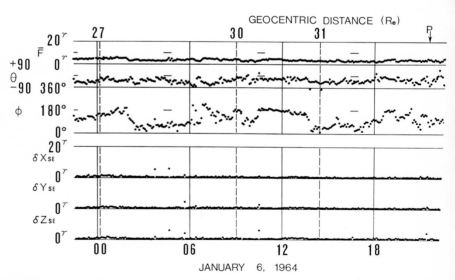

JANUARY 6, 1964

Figure 5.12. a. Magnetic field data from orbit 11 on January 5, 1964. The out-bound orbit passes out of the magnetic cavity into the magnetopause at $13.6 R_e$ and through the shock wave into interplanetary space at $19.7 R_e$. The variables are defined in Fig. 5.11.

b. Magnetic field data from orbit 11, January 6, 1964. The spacecraft con-tinuously measures the magnetic field in interplanetary space. Both graphs from Ness, Scearce, and Seek,[37] 1964.

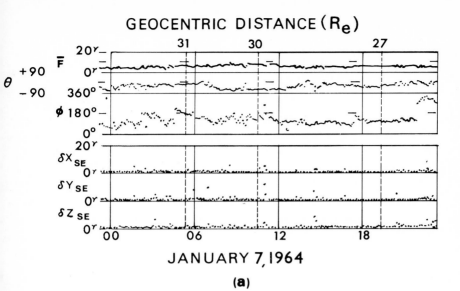

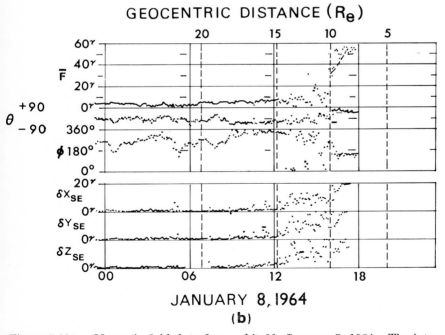

Figure 5.13. a. Magnetic field data from orbit 11, January 7, 1964. The interplanetary field is continuously measured in interplanetary space. A null surface where ϕ changes by 180 deg is apparent near the right edge of the graph. The variables are defined in Fig. 5.11.

b. Magnetic field data from orbit 11 on January 8, 1964. The spacecraft goes from interplanetary space through the shock wave into the magnetopause at $16.0 R_e$ and from the magnetopause into the magnetic cavity at $9.7 R_e$. Both graphs from Ness, Scearce and Seek,[37] 1964.

along the top. The average magnetic field, the latitude and longitude of the magnetic field directions in the solar ecliptic coordinate system and the variances are plotted.

The data in Fig. 5.12, for January 5, shows the Imp 1 spacecraft leaving the magnetic cavity at 13.6 R_e. There is an abrupt decrease in the average magnetic field $\overline{F}$. A slight change in the direction of the magnetic field perpendicular to the ecliptic plane and a large change in the direction in the ecliptic plane is observed at the boundary. Farther out the spacecraft crosses the shock boundary at 19.7 R_e. In the interval between 13.6 and 19.7R_e the magnetic field is variable both in magnitude and direction. Beyond the shock boundary in inter-planetary space the magnetic field approaches a constant value of a few gamma. The directions of the magnetic field still vary some but the variances are greatly reduced.

This constant magnetic field pattern continues through January 6 and 7 in Figs. 5.12 and 5.13 until 2200 on January 7 at which time there is a reversal by 180 degrees in the direction of the magnetic field in the ecliptic plane. The authors interpret this as the satellite passage through a neutral plane in the magnetic field. As the spacecraft continues beyond apogee and reenters the shock region and the earth's magnetic cavity the magnetic field pattern repeats. However, this time the boundaries are passed at 16.0 and 9.7R_e as the reentry is much nearer the sun–earth line than on the trip out.

The magnetic fields in interplanetary space were usually quite constant from 4 to 7 γ with variances that were usually less than 0.4 γ. There was a strong preference for the magnetic field to have a direction of 40 to 50 deg with the earth–sun line as expected theoretically. The upstream angle which differs by 180 deg was also often found. The distribution of the directions of the magnetic field in the ecliptic plane and with respect to the normal to the ecliptic plane are shown in Fig. 5.14.[38] The positive fields, directions away from the sun, occurred 48 per cent of the time and the negative fields, directions toward the sun, 35 per cent. In between, the interplanetary magnetic fields decreased abruptly to zero at the null surfaces separating the regions where the magnetic fields were in opposite directions.

The interplanetary magnetic field appeared to prefer directions below the plane of the ecliptic by 10 to 20 deg. Such a magnetic field that points out of the plane of the ecliptic is very difficult to explain theoretically. The theory permits no magnetic fields perpendicular to

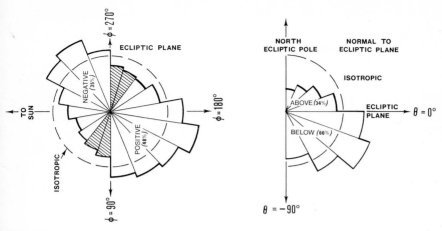

Figure 5.14. The number of interplanetary magnetic field directions/ster in the ecliptic plane and perpendicular to the ecliptic plane.   The dashed curves correspond to isotropic distributions.   Positive is away from the sun and negative is toward the sun.   The distribution in the ecliptic plane has its maximum at about 50 deg from the sun–earth line.   The distribution in angle perpendicular to the ecliptic plane shows that the interplanetary field is predominately parallel to the ecliptic plane.   From Wilcox and Ness,[43] 1965.

the sun's equator except for fluctuations that average out to zero over a period of time.   Since the plane of the equator of the sun makes an angle of 7 deg with the plane of the ecliptic, the maximum magnetic field angle allowed out of the ecliptic is 7 deg.   Over the course of one year this angle will average to 0 deg.

The first solid evidence for the curved magnetic field lines did not come from satellite experiments but rather from cosmic-ray measurements on the surface of the earth.   Ken McCracken, in 1962, in a beautiful analysis[39] of the solar proton times of travel from the sun to the earth and their directions of arrival deduced the direction of the interplanetary magnetic field.   He observed that of the 14 solar flares detected on the surface of the earth from 1942 to 1961, only 1 came from the eastern 40 per cent of the solar disk.   There was a clear tendency for the short rise times from 20 min to one hour to occur on the western limb of the sun.   Flares that arose near the center of the sun had rise times from one to several hours and the one flare observed on the far eastern limb had a rise time of 5 hours.   He attributed these times to the properties of interplanetary space and not to differences of the sources on the surface of the sun.

McCracken's picture of the interplanetary magnetic field is shown in Fig. 4.27. The active sunspots on the western limb of the sun are connected to the earth through interplanetary magnetic field lines that make angles of about 50 deg with the sun-earth line. Energetic solar protons of several hundred MeV injected onto the magnetic field lines near the sun have pitch angles of less than 1 deg when they arrive at the earth. This is true if each conserves the first adiabatic invariant, its magnetic moment, in the interplanetary magnetic field that decreases like $1/r^2$ from the sun. This directional flux of protons then strikes small areas on the earth called "impact zones." The travel times and observed rise-times should be short. These effects were observed in the flares of May 4 and November 15, 1960. And the maximum fluxes arrived from an angle of about 50 deg from the sun–earth line.

One additional ingredient was needed in the model, small-scale irregularities in the interplanetary magnetic field. The irregularities were necessary to explain the solar protons with angles up to 80 deg in the May 4, 1960 solar flare. Irregularities in the interplanetary magnetic field of sizes comparable to the cyclotron radii of the protons can scatter the protons and increase their pitch angles. In that case the first adiabatic invariant is no longer conserved. The radius of curvature of a 500 MeV proton in a magnetic field of 2.5 γ is 0.01 AU. To explain the observed average pitch angle of 45 deg requires a total scattering of about 90 deg. This also compensates for the collimation caused by the decreasing magnetic field. The average scattering angle after η encounters of ϵ deg each is $\sqrt{\eta}\,\epsilon$. Taking ϵ equal to 15 deg as suggested by the magnetic field data,[33] we find that 36 encounters are required. This number of irregularities appears reasonable for the picture of the interplanetary magnetic field given in Fig. 4.27.

The interplanetary magnetic field also traps the solar protons and excludes the galactic cosmic rays. This produces the Forbush decrease in the galactic cosmic ray intensity. The Forbush decrease is greatest at times of solar storms and at the maximum of the solar cycle.

From the Imp 1 magnetic field experiment, Norman Ness and John Wilcox[38] discovered that the interplanetary field changes with the sun's longitude. It changes directions toward or away from the sun in alternate sectors of the giant wheel that lies in the plane of the ecliptic. In the time period of December 1963 to February 1964, the magnetic field pointed away from the sun in each of two sectors that formed 2/7 of the total circle. Each sector with the field pointing away from the sun was separated by one with the field pointing toward the sun. One sector away from the sun was 2/7 and one was 1/7 of the total circle. The sector structure rotates with the sun and sweeps past the earth every 27 days. The neutral sheet between the boundaries of the sectors is very thin and the fields change directions in a few minutes.

The sector wheel is shown in Fig. 5.15. The plus signs indicate magnetic fields away from the sun and minus signs toward the sun. These are three-hour averages of the magnetic field. The directions are consistently plus or minus in each of the sectors after the magnetic storm on December 2 until the end of data acquisition 3 orbits later.

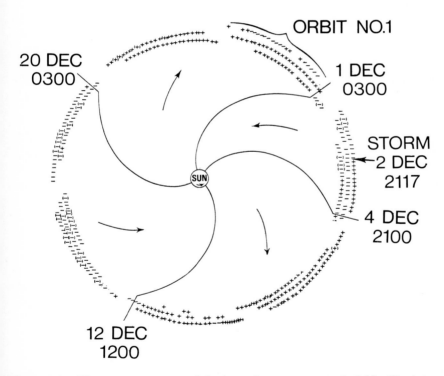

Figure 5.15. The sector structure of the interplanetary magnetic field. The inter-planetary magnetic field is divided into four sectors of 2/7, 2/7, 2/7, and 1/7 of the arc of a circle. The magnetic field alternates between directions away from the the sun +, and toward the sun −. Data from 3 successive orbits are plotted. Shortly before the start of orbit 1, a magnetic storm occurred, and the data there varies considerably. From Wilcox and Ness,[43] 1965.

The magnitude of the magnetic field, the solar wind density, the solar wind velocity, and the geomagnetic activity were studied within a sector.[38] The magnetic field varied from about 3.5 to 6.5 γ, the solar wind velocity from 280 to 340 km/sec, the solar wind density from 7 to 14 protons/cm³, and the geomagnetic activity index K_p from 10 to 25. The magnetic field and solar wind velocity measured on Imp 1 and the geomagnetic activity measured on the surface of the earth all varied in the same way across a 2/7 sector. They peaked at the second day and

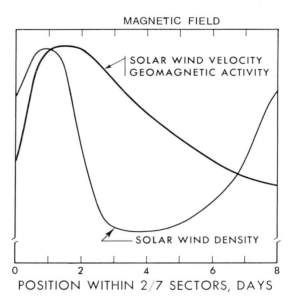

Figure 5.16. Plot of the average interplanetary magnetic field, solar wind velocity, solar wind density, and geomagnetic activity versus position within a 2/7 sector. The magnetic field varies from 3.5 to 6.5 γ, the velocity from 280 to 340 km/sec, the density from 7 to 14 protons/cm³ and the 24 hour sum of the geomagnetic activity index $\sum K_p$, from 10 to 25. From Wilcox,[40] 1966.

decreased to the end of the sector on eighth day. The solar wind density measured on Imp 1 behaved differently. It peaked on the first day, dropped to a minimum on the fourth day and increased again to the end of the sector. These variations are given in Fig. 5.16.[40] The high velocity solar wind carrying an increased magnetic field seems to be responsible for the magnetic activity on the surface of the earth.

Data from the magnetometer on Imp 2[41] from October 4 to November 30, 1964 were quite similar to data from Imp 1 about one year earlier. However, the sectors were more equal in size. In each sector the ratio of the number of measurements of the magnetic field in the sector direction to the number in the opposite direction was 100/1.

Imp 1 also carried a solid-state detector that measured the flux of protons of a few MeV continuously during three solar rotations.[42] C. Fan, G. Gloeckler and J. Simpson from the University of Chicago found protons at one time only during the 27-day solar rotation period. The protons always appeared during the same magnetic solar sector[43] and were detected on the satellite only when the sun's source of protons was connected to the satellite by magnetic field lines.

5.4 SOLAR-PROTON STREAMS

The Explorer 12 satellite was launched in August, 1961, with an apogee of $13R_e$ on the sunlit side of the earth. The proton experiment of D. Bryant, T. Cline, U. Desai and F. McDonald, from NASA-Goddard[44] was in a position to observe the solar-proton streams outside the earth's magnetic cavity but inside the shock front located at $14R_e$. On September 30, 1961, two days after a class-3 flare, a stream of protons was detected with energies of a few MeV and of intensity 10 times higher than in the flare. A magnetic storm with its sudden commencement and auroral displays at mid-latitudes was coincident with the proton stream. A Forbush decrease of the cosmic ray protons was detected both by a cosmic ray detector on Explorer 12 and by neutron monitors on the earth. The proton intensity decreased to its background level by October 7 but then increased again to a high intensity on October 27. This was just 27 days, one solar rotation, after the original increase. And the protons were again accompanied by a magnetic storm and Forbush decrease. The Explorer 12 experiment also detected solar protons on two additional occasions.

The recurrence of magnetic storms with the 27-day period of the sun's rotation had previously led Julius Bartels[45] in 1932 to suggest that streams of plasma were continuously emitted from long-lived sources on the sun. He called these regions M-regions.

The Chicago proton detector on Imp 1[42] was especially well designed to measure the energies of protons at the low fluxes found in inter-

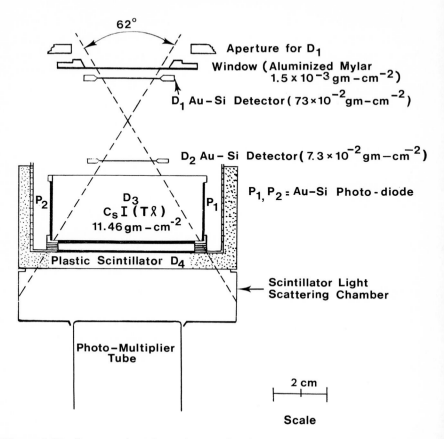

Figure 5.17. Cross-section through the absorbers and detectors of the Imp 1 cosmic-ray telescope. Surface barrier detectors D_1 and D_2 and the CsI scintillator D_3 make up the telescope. The photodiodes P_1 and P_2 detect the light from the CsI scintillator. The plastic scintillator cup D_4 is placed in anticoincidence with D_1, D_2 and D_3. Pulses from D_1 and D_3 are analyzed. From C. Y. Fan, G. Gloeckler and J. A. Simpson, "Cosmic Radiation Helium Spectrum below 90 MeV per Nucleon Measured on Imp 1 Satellite," *J. Geophys. Res.* **70**, 3515–3527 (1965).

planetary space. It consists of two surface barrier detectors D_1 and D_2 in coincidence with the signal from the following CsI detector D_3 as shown in Fig. 5.17. An anticoincidence plastic scintillator cup D_4 around the CsI detector eliminates the particles that escape from the CsI detector. If a coincidence occurs between D_1 and D_2, the signal from D_1 goes to a pulse height analyzer. If the particle penetrates to

D_3, that signal is also analyzed. This enables the telescope to identify and measure the energies of protons, α-particles, and heavier ions.

With this detector protons with energies greater than 1 MeV were detected for six consecutive 27-day intervals. The detector was modified to measure lower energy protons and α-particles and was flown on the OGO-1 satellite in late 1964. Low energy α-particles were detected that correlate in time with changes in proton energy and fluxes. They appear to have the same origin as the streaming protons.

Additional evidence for the long-term persistence of the solar streams came from the NASA-Goddard cosmic-ray experiment[46] on Explorer 14 for the time period of February to July 1963. Protons of 3 to 20 MeV were observed on seven consecutive rotations of the sun. The proton beam widths were 30 to 120 deg wide at the earth and protons of different energies all arrived at the same time. This absence of time dispersion, the steep energy distributions, and the low intensities are convincing evidence for equilibrium proton streams rather than for bursts of solar protons from solar flares.

By comparing the measurements from proton detectors on the spacecraft Imp 3 and on Mariner 4,[47] it was possible for J. O'Gallagher and J. Simpson to find the times that the proton streams reached the different positions in interplanetary space. The proton increase on day 239, 1965 at Mariner 4 was seen 4.7 days later at Imp 3 near the earth. The predicted rotation time was 4.0 days. Likewise, the proton increase on day 265 at Mariner 4 was seen 5.2 days later near the earth in agreement with the predicted rotation time of 5.2 days. In addition, the Mariner 4 magnetometer showed reversals of the magnetic field direction from plus to minus as expected from the measured changes in the proton fluxes.

Evidence[48] that there are kinks in the Archimedes spiral magnetic field lines was found when the directions of energetic solar protons were compared with the directions of the magnetic field. In Fig. 5.18 the direction of the magnetic field is given by a line with a black dot at the end. The solar proton direction is given by the arrow. The two directions are very close even during times of large changes of directions. If the magnetic field vectors are added the large kink results. It is associated with the magnetic field line from the sun to the earth shown in Fig. 5.19. In this picture, the magnetic field lines from the sun are sometimes smooth Archimedes spiral curves and sometimes kinked.

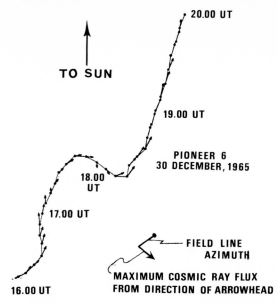

Figure 5.18. The directions of the magnetic field in the ecliptic plane and the maxima of the cosmic ray fluxes from 1600 to 2000 UT, December 30, 1965. A dot followed by a line gives the direction of the magnetic field. The direction of the maximum cosmic ray flux is opposite the direction of the arrow. The two directions are always nearly the same and both change in the same way during the kink at 1800 UT. From McCracken and Ness,[48] 1966.

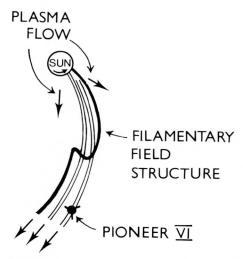

Figure 5.19. Interplanetary magnetic field filaments. The magnetic field filaments are sometimes smooth Archimedes spirals and sometimes kinked. The kinked filament was derived from the measurements of Fig. 5.18. From McCracken and Ness,[48] 1966.

In spite of the kinks, filaments of the magnetic field retain their relative position in longitude to one another.

The solar proton measurements of Ken McCracken, U. R. Rao and R. P. Bukata on Pioneer 6[49] suggested that the sector structure observed early in 1964 was essentially unchanged in early 1966. They proposed a model for interplanetary space to explain the sector structure of the magnetic field based on Eugene Parker's shock theory.[19] Standing shock waves define the boundaries of the magnetic field sectors. The fast plasma from a hot spot in the corona overtakes the slower plasma from the normal solar wind and a standing shock is formed as shown in Fig. 5.20. The increased plasma density in the shock is responsible

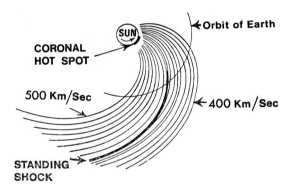

Figure 5.20. Schematic drawing of the formation of a standing shock. The shock is formed at the interface between the normal solar wind of 400 km/sec and a faster solar wind from a corona hot spot of 500 km/sec. Such standing shock waves act like the vanes on a centrifugal pump and may scatter cosmic ray protons and α-particles and cause the Forbush decreases. From McCracken, Rao and Bukata,[49] 1966.

for the higher geomagnetic activity while the increased magnetic field strength within the shock inhibits cosmic-ray diffusion across the magnetic field lines. The standing shock waves permanently rotate with the sun like the vanes of a huge centrifugal pump. Each standing shock wave sweeps across the earth every 27 days. The number of cosmic ray protons in interplanetary space decrease by deflecting off these vanes.

5.5 SOLAR WIND TEMPERATURE

Superimposed on the stream velocity of the solar wind are the lower thermal velocities. The thermal velocity in any direction is $v_t = v - v_s$ where v is the total velocity in that direction and v_s is the stream velocity. The number of protons with v_t is

$$n(v_t) = n_0 \, e^{-1/2}\left(\frac{m_p \, v_t^2}{kT}\right) \qquad (5.27)$$

where k is Boltzman's constant, m_p is the proton mass, T is the temperature of the protons and n_0 is a constant. Marcia Neugebauer and Conway Snyder[29] fitted the shapes of the Mariner 2 solar wind velocity peaks to the Maxwell–Boltzman distributions and derived temperatures. A broad peak signified a high temperature and a narrow peak a low one. The average daily temperature was between 1.51 and 1.85×10^5 °K. This is a factor of 10 lower than the temperature at the base of the corona.

The highest temperatures were found in the high-velocity streams. These were a factor of 5 higher than the average temperatures. The highest 3-hour average temperature was 8.0×10^5 °K. Between the streams the plasma was much cooler with a minimum 3-hour average temperature of 3×10^4 °K.

The ratio ξ of the proton thermal energy density to the proton stream energy density is

$$\xi = \tfrac{3}{2}kT / \tfrac{1}{2}m_p v_s^2 \qquad (5.28)$$

The value ξ was always quite low with a maximum 3-hour average of 0.53 and a minimum value too low to be measured. And it appeared that the thermal energy density was usually about as large as the magnetic field energy density.

It was not clear whether the temperature was caused by the thermal-motion conducted up from the source of protons at the bottom of the corona or whether the protons were heated in interplanetary space by hydromagnetic wave interactions.

The velocity distributions from the curved-plate electrostatic analyzers of J. Wolfe, R. Silva, D. McKibbin and R. Mason from NASA-Ames[50] on Pioneer 6 were found to be quite anisotropic. The temperature along the magnetic field direction was several times larger than the temperature perpendicular to the magnetic field direction.

The Vela 3 results of A. Hundhausen, J. Asbridge, S. Bame, H. Gilbert and I. Strong[28] from Los Alamos definitely show that the solar wind random motion is not thermal—it is not even isotropic. The angle and energy resolutions of the electrostatic analyzer on Vela 2 and 3 were sufficiently good that velocities in two perpendicular directions could be measured. They were the direction of the sun–satellite line v_1 and the direction perpendicular to that line and to the spin axis of the satellite v_2. The spin axis made an angle of about 60 deg to the ecliptic plane. Contour plots of constant proton flux in this v_1v_2 plane were not circular but rather showed an elongation in one direction as can be seen from Fig. 5.21.

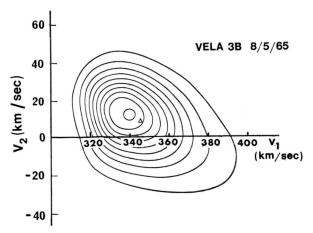

Figure 5.21. Contours of constant velocity in the v_1 v_2 plane measured by the Vela 3 electrostatic analyzer. v_1 is the velocity in a direction perpendicular to the sun–earth line and to the satellite spin axis. The highest velocity occurs at the circle in the center of the diagram. The stream velocity is plotted as a triangle also near the center of the diagram. An elongation occurs in the direction of the magnetic field lines. Interplanetary plasma temperatures may be derived from this diagram. From Hundhausen, Asbridge, Bame, Gilbert and Strong,[28] 1967.

The stream velocity is indicated with a triangle. The highest flux is located at the circle in the center of the contours close to the triangle. The values of v_t^2 were calculated from Fig. 5.21 and plotted versus the angle to the sun direction. The resultant temperature plot was usually elongated in a direction of about 50 deg to the sun–earth axis in a direction away from the sun. The directions of the axes of these temperature diagrams Φ_T were plotted on the polar diagram of Fig. 5.22.

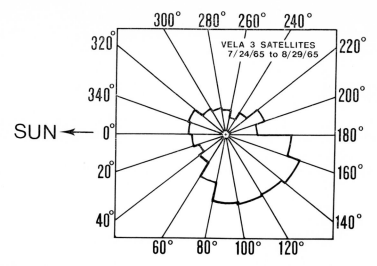

Figure 5.22. The distribution of the angles in the v_1v_2 plane where the temperature, or random energy is the highest. See Fig. 5.21. The area of each 20 deg wedge is proportional to the number of times that the angle fell within the interval. The direction of the maximum of this distribution is the same as the maximum of the magnetic field direction distribution in the ecliptic plane of Fig. 5.14. From Hundhausen, Asbridge, Bame, Gilbert and Strong,[28] 1967.

The distribution of these angles has a primary maximum at 130 deg and a smaller secondary maximum at 330 deg. The random velocities are definitely not symmetric but have their maximum at an angle of 50 deg to the sun–earth direction away from the sun. The ratio of the maximum to the average temperature varies from 1.0 to 2.5 with an average value of 1.4.

The random velocity distribution of Fig. 5.22 looks very much like the magnetic field direction distribution of Fig. 5.14. The average direction of the highest random energy of the protons is in the same direction as the magnetic field. The random energy is lowest in a direction perpendicular to the magnetic field.

A severe anisotropic distribution of the random velocities is expected at the earth's orbit. From the conservation of the proton's magnetic moments in their motion from the sun to the earth we have

$$\frac{B_\odot}{B_e} = \frac{\sin^2\alpha_\odot}{\sin^2\alpha_e} \tag{5.29}$$

The $\odot$ and e refer to the sun and the earth respectively, and the pitch angle α is the angle between the direction of the proton's velocity and the direction of the magnetic field. Since the magnetic field of the sun falls off roughly as one over the distance squared and since the distance from the sun to the earth is $200\,R_\odot$, we expect the pitch angle to decrease by a factor of about 200 in coming from the sun to the earth. If the distribution in random velocities were isotropic at the source, the particles should be collimated when they reach the earth with a large peak in the forward direction along the magnetic field.

The question is now turned around. Instead of asking why the random velocities are peaked in the direction of the magnetic field, we ask why they are not more peaked along the magnetic field direction? Why isn't the energy along the magnetic field direction much greater than the energy perpendicular to the magnetic field direction? There are at least two possible ways of decreasing the anisotropy. Collisions among the ions will redistribute the energy from parallel to the magnetic field, to perpendicular to it. Instabilities also develop when $E_{\parallel}/E_{\perp}$ is greater than a limiting value that depends upon the proton and magnetic field energy densities. The instabilities lead to the growth of waves and particle-wave interactions.

The above measurements imply that the direction of the maximum temperature is instantaneously aligned along the direction of the maximum magnetic field. However, simultaneous measurements of the maximum temperature direction and maximum magnetic field direction must be made to finally settle the question.

5.6 SHOCK WAVES

Fluctuations should occur in the magnetic field, the plasma density and the temperature of interplanetary space. Eugene Parker[19] discussed the fluctuations and the propagation of the waves and suggested sources for the fluctuations. The inner corona rotates with the sun. Since the hotter regions of the corona expand more rapidly than the cooler regions, they overtake the cooler regions in interplanetary space. The shocks and turbulent regions sketched in Fig. 5.23 can be formed. The compression and rarefaction regions form curves like the vanes on a centrifugal pump and shock waves form at the compression boundaries.

Fluctuations in the solar wind occur when the corona temperature varies with time. A sudden increase in coronal temperature following a large solar flare creates blast waves propagating outward from the sun

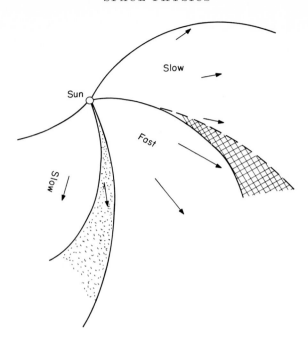

Figure 5.23. Sketch of the compression (cross hatched) and rarefaction (dotted) regions in interplanetary space. A shock front or blast wave is formed at the interface where the faster solar wind from a corona hot spot catches up with the slower, normal solar wind. This blast wave may be responsible for the scattering of cosmic ray protons and α-particles and for the Forbush decreases. See Fig. 5.20. From Parker,[19] 1965.

through the solar wind. The blast wave is formed by scooping up the material in the quiet-day wind ahead of the shock. In one extreme case, the heated corona pushes so hard that the velocity of the wave is kept constant; in another, the pressure on the back of the wave is zero so that the energy of the wave is constant and the velocity falls off like $1/r^{1/2}$.

Small-scale disturbances are generated by the transitions between the slow and the fast solar wind velocities. They are also generated by the fluctuations in the lower corona and the motions that are responsible for heating the corona. The fractional disturbance in the magnetic field at the sun is amplified in interplanetary space. A small fractional amplitude of $dB_\odot/B_\odot = 0.05$ at the sun grows to $dB_e/B_e = 1$ at the orbit of the earth. In this way an observer at the orbit of the earth listens to the disturbances that take place on the surface of the sun.

The first example of a magnetohydrodynamic shock in interplanetary space was found by the NASA-Ames magnetic field and plasma experiments on Mariner 2.[51] The change in the plasma density and magnetic field occurred on October 7, 1962. The magnetic field rose from 6 to 16 γ in less than 3.7 min., immediately fell to 11 γ and continued irregular for many hours. A sudden commencement magnetic storm was observed on the surface of the earth 4.7 hours later. A spherical wave that originates in the sun takes just this time to travel from the satellite to the earth at a velocity of 510 km/sec. Since the stream velocity of the solar wind after the shock was 460 km/sec, the wave velocity relative to the solar wind was only 50 km/sec. The preshock temperature was found to be 1.2×10^5 °K and the postshock temperature 1.7×10^5 °K.

A sudden increase in the solar wind density by a factor of five was measured on the Vela 2 satellites[52] during the large magnetic storm of April 17–18, 1965. The sudden change in the solar wind density occurred at about the same time as the sudden magnetic impulse on the earth at 0900 UT April 18. Within 6 minutes the solar wind velocity changed abruptly from 405 to 415 km/sec, the temperature jumped from 6×10^4 °K to 1.3×10^5 °K and the proton density rose from 5 to 28 protons/cm³. After the passage of the interplanetary shock the stream velocity increased to 550 km/sec, the temperature to 7×10^5 °K and the proton density decreased to a constant level of 11 protons/cm³.

Two cases of interplanetary shock waves moving through the solar wind were detected by the electrostatic analyzer experiments on the Vela 3 satellites.[53] The mean travel velocity for shock waves from the sun to the earth was obtained by dividing the sun–earth distance by the time interval between the flare origin and the disturbance arrival. In each case the shock wave velocity past the spacecraft was much less than the mean shock wave velocity.

In the January 20, 1966 case, the velocity increased from 340 to 385 km/sec across the shock, the proton density increased from 9 to 23 protons/cm³ and the average temperature rose from 3×10^4 to 8×10^4 °K. These changes are shown in Fig. 5.24. The shock velocity past the satellite was calculated to be 410 km/sec which resulted in a shock velocity through the solar wind of 70 km/sec. A 2B solar flare occurred on the sun at 2253 UT, January 18. It was located at 7 deg east of the sun–earth line and was the source of 5 MeV protons at the earth's orbit. If this flare were the source of the shock the average

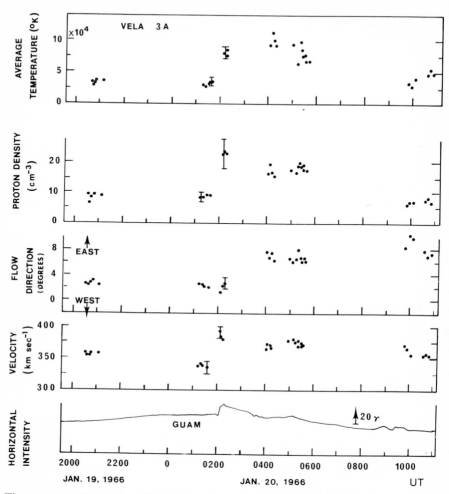

Figure 5.24. Plot of the solar wind proton velocity, stream direction, density and temperature for January 19 and 20, 1966. The abrupt changes in these quantities at 0200 UT, January 20 are compared to the change in the horizontal magnetic field measured at Guam. The correlation among all 5 quantities is excellent. From Gosling, Asbridge, Bame, Hundhausen and Strong,[53] 1968.

transit velocity of the shock wave to the earth would be 1670 km/sec. An earlier 2B solar flare at E27 at 1032 UT on January 17 could possibly have been the source but it did not produce a measurable flux of 5 MeV protons. If this flare were responsible for the shock the average transit velocity would have been 686 km/sec. In either case the transit velocity is significantly greater than the velocity of the shock past the satellite.   A second similar shock occurred on October 5, 1965 where the mean transit velocity of the disturbance from the sun also greatly exceeded the velocity of the shock wave past the satellite.

In each of the shocks the velocity through the solar wind was about 80 km/sec, 415 km/sec past the spacecraft.   Since the velocity of the shock at the spacecraft was much less than the average transit velocity through interplanetary space, the authors concluded that the shock waves were slowed down as they propagated through the solar wind. The decrease in velocity is the result of the expansion of the shock wave and of the transfer of energy to the solar wind by heating and by acceleration of the plasma in the stream direction.

Large changes in temperature are expected across shock fronts and this is indeed measured for the earth's bow shock near the earth's magnetic cavity, see Chapter 6.   Furthermore, continuity across the shock front demands that the change in the velocity perpendicular to the shock front be a sizeable fraction of the shock velocity in the solar wind. Since these large changes have not been observed in the almost continuous measurements since 1962, the authors conclude that the shock waves from the sun are significantly decelerated by the time they reach the orbit of the earth.

5.7 MAGNETIC FIELD POWER DISTRIBUTION

The variation of the magnetic field power with frequency was calculated from the data of the Mariner 2 magnetometer experiment.[54]   The magnetic field variations were analyzed at frequencies between 1.16 and 1160 cycles/day.   The highest frequency was determined by the sampling rate and the lowest frequency by the fact that most of the magnetic field power was contained above that frequency.   The exceptions were the reversals of the magnetic field directions because of the 27-day rotation of the sun.   The total power contained in each of the three spherical components of the magnetic field variations was $P_r = 4\gamma^2$, $P_\theta = 8\gamma^2$, and $P_\phi = 6\gamma^2$.

The composite power spectrum for B is given in Fig. 5.25.   The distribution varies with frequency as $1/f^{1.5}$ at low frequencies and as $1/f$ at the high frequencies.   The power levels for B are significantly lower than the values expected from purely random fields.   This suggests that the interplanetary magnetic fields are partially organized and that

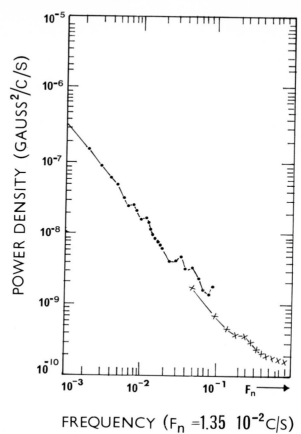

Figure 5.25. The magnetic field power density versus frequency measured on Mariner 2 for the days 270 to 279 and 282 to 291, 1962. The horizontal axis is in units of F_n where $F_n = 1.35 \times 10^{-2}$ c/s. The power density falls off approximately like $1/f$. From Coleman,[54] 1966.

coherent variations exist between the interplanetary magnetic fields and the solar wind velocity. In addition, the power distributions show that the magnetic field disturbances in the radial direction are weaker than in the polar or azimuthal directions.

With the magnetometer on Mariner 4,[55] the fluctuations of the magnetic field power were studied over shorter time periods of 1 sec to 1 hour—frequencies of 3×10^{-4} to 0.5 c/s. The data were obtained toward the end of 1964 during the minimum of the solar cycle. Power distributions were analyzed for quiet, intermediate and active days.

The power densities were sometimes 15 times as great on the active as on the very quiet days.

The observed discontinuities in the magnetic field were divided into shock waves or contact surfaces following the review article of D. S. Colburn and Charles Sonett.[56] Very few shock waves were observed by this experiment but many contact surfaces were found. A contact surface is identified by a change in magnetic field strength or direction, a change of plasma density or composition, or a change in plasma stream velocity. A 4 γ magnetic field change was required. In over half of the time this change occurred in a thickness of less than 3500 km and in 10 per cent of the time in less than 500 km.

Simple discontinuities occur by a rotation of the magnetic field from its initial to its final direction. These are called tangential discontinuities.[56] They are formed at filament boundaries separating regions of solar plasma and of magnetic fields which have different times of origin on the sun. The filament transitions have two clear discontinuities, one on each side of the filament, and the magnetic fields on both sides often have the same direction. The time between the two boundaries varies from less than one minute to more than one hour. The filament boundaries are usually nearly parallel with a highly elliptical cross section and the tubes of magnetic field and plasma probably extend back to the sun.

5.8 COSMIC RAY MODULATION BY INTERPLANETARY MAGNETIC FIELDS

The cosmic ray intensity decreases during the maximum of the sunspot cycle and during magnetic storms. These decreases are called Forbush decreases after Scott Forbush, their discoverer. The theory to explain these decreases in the cosmic ray intensities in the magnetic fields of interplanetary space was developed by Eugene Parker in his book *Interplanetary Dynamical Processes.*[57] There he describes his theory of the solar wind and applies it to the modulation of cosmic rays and solar protons in interplanetary space.

The cosmic ray charged particles are convected outward from the inner solar system by interactions with the magnetic fields that are continuously carried outward by the solar wind. The Forbush decreases during magnetic storms are caused by blast waves that follow solar flares. The Forbush decrease during the solar cycle is is caused by the

cumulative effects of the outward notions of the large scale interplanetary fields that probably extend out to 10 to 100 AU. And the
outward convection of the cosmic ray particles is balanced by their
inward diffusion and drift.

In one model to explain the Forbush decrease during magnetic storms,
Parker calculated the amount of diffusion of the cosmic rays through
the disordered magnetic fields of expanding plasma clouds in interplanetary space. He found that this diffusion was inefficient for scattering the cosmic rays and that it took disordered magnetic fields of
10^{-3} gauss at the orbit of the earth to produce the observed Forbush
decreases. Nor was it known how to manufacture the stable disordered
clouds.

In his blast wave model to explain these same sudden Forbush
decreases during solar flares, quiet-day interplanetary magnetic fields
are swept up and compressed into the blast wave that originates in the
lower corona. The rise time for the cosmic ray decrease is the transit
time of the wave from the sun to the earth plus its time to pass the
earth. This may take from a few hours to a day or so. The recovery
time for the blast wave depression takes from one to several days.

To explain the general depression of the cosmic rays during the
maximum of the solar cycle, Parker uses the Fokker–Planck diffusion
equation

$$\frac{\partial n_{\mathrm{cr}}}{\partial t} = -\mathbf{\nabla} \cdot (\boldsymbol{v}_s \, n_{\mathrm{cr}}) + K\nabla^2 n_{\mathrm{cr}} \tag{5.30}$$

where n_{cr} is the cosmic ray density, v_s is the solar wind velocity and K
is the diffusion coefficient

$$K = v_{\mathrm{cr}}\lambda_{\mathrm{cr}}/3 \tag{5.31}$$

The v_{cr} is the cosmic ray velocity and λ_{cr} is the diffusion mean free path.
The diffusion arises from the disordered magnetic fields caused by the
degraded blast waves and nonuniformities in the solar wind. Taking
spherical symmetry about the sun, and the time independent case
$\partial n/\partial t = 0$ he finds for the cosmic ray density

$$n(r, E) = n(\infty, E) \exp\left(-\int_r^{\mathscr{L}} v_s \, dr/K \right) \tag{5.32}$$

where $n(r, E)$ is the cosmic ray density outside the solar wind cavity
and $\mathscr{L}$ is the effective modulation distance. The exponent increases by
1 or 2 from minimum to maximum of the solar cycle for cosmic ray

protons of 1 BeV. The characteristic time for the build-up and relaxation is the quiet-day transit time, about 1.6 years, the time for the solar wind to travel 100 AU.

In the diffusion theory the anisotropic diffusion of cosmic rays is caused by the fluctuating component of the magnetic field that scatters the cosmic rays. The balance between the outward convection of the cosmic ray particles by the solar wind and the inward diffusion produces the radial gradient in the cosmic ray density. This gradient causes the 11-year solar cycle and the 27-day solar modulations of the cosmic ray intensity. W. Axford[58] suggested that the density of cosmic rays in the galaxy should be larger than near the earth. The velocity of the solar wind should also be substantially lower at large distances from the sun because of the work done in pushing the cosmic ray particles away from the sun. This decreases the size of the interplanetary solar wind cavity produced by the interaction of the solar wind with the interstellar medium.

The diffusion coefficient parallel to the magnetic field direction is the same as before, see Eq. 5.31, and the diffusion coefficient perpendicular to the magnetic field direction is

$$K_\perp = \frac{1}{3} \frac{v_{cr} \lambda_{cr}}{1 + (\omega_B \tau)^2} \tag{5.33}$$

The τ is the mean time between collisions and the cyclotron frequency ω_B is usually so large that $\omega_B \tau \gg 1$ and then $K_\parallel \gg K_\perp$. The model incorporates a radial supersonic wind that flows out to a boundary at 10 to 100 AU then becomes subsonic and disperses into interstellar space. The usual spiral magnetic field was used inside the boundary and a highly irregular magnetic field with $\omega_B \tau \ll 1$ outside the boundary. The gradient in the cosmic ray density was found to be

$$\frac{dn}{dr} \simeq \frac{nv_s}{K_\parallel \sin^2 \zeta} \tag{5.34}$$

where ζ is the angle between the magnetic field and the direction to the sun. The cosmic ray stream direction is parallel to the orbital motion of the earth about the sun and its magnitude is equal to the velocity of rotation of the magnetic field lines around the sun. The cosmic ray gradient is positive so increases with distance from the sun. For a solar wind velocity of 500 km/sec and $\zeta = 45$ deg, he finds $(dn/dr)/n = 0.1/\text{AU}$. This radial density gradient and the changes in radius of the solar wind cavity R_s and in λ_{cr} account for the changes in the cosmic ray intensity over the solar cycle.

Above some energy like 10 BeV the shape of the cosmic ray spectrum is independent of the time in the solar cycle. The particles are unaffected by the fluctuations in the interplanetary magnetic fields and λ_{cr} is very long compared to R_s. For intermediate energies the modulation is important only during the maximum of the solar cycle but for the lowest energies the modulation is effective throughout.

G. Gloeckler and J. Jokipii[59],[60] evaluated the diffusion coefficient from the power spectrum of magnetic irregularities observed on Mariners 2[54] and 4.[55] They found that the diffusion coefficient $K_{\parallel}$ is proportional to Rv_{cr} where R is the rigidity of the particles. Such a variation with rigidity is required by the observed changes of the cosmic ray protons and alpha-particles over the solar cycle. The energy distributions of the alpha-particles measured on Imps 1, 2 and 3 are shown in Fig. 5.26 for three different times during a two year period just before

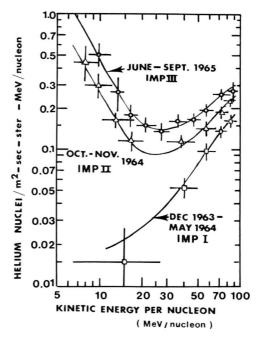

Figure 5.26. α-particle differential energy distributions for three times near the minimum of the solar cycle. The low energy α-particles increase as the solar cycle minimum is approached. From Gloeckler and Jokipii,[59] 1966.

the minimum of the solar cycle.[61] For energies above 30 MeV/nucleon the helium flux increases as the minimum of the solar cycle is approached. Below 30 MeV/nucleon there is a spectacular increase of of a factor of 10.

The cosmic ray particles with cyclotron radii r_B are mostly scattered by the fluctuating magnetic fields perpendicular to the solar wind velocity at frequencies near

$$f_0 = v_s/2\pi r_B \qquad (5.35)$$

The Mariner 2 and 4 quiet time magnetic power spectra vary approximately as $P(f) = \delta/f$ where $\delta = 0.7 \times 10^{-10}$ gauss^2/(cycle/sec) and then

$$K_{||} = \frac{B_0}{3\pi\delta} Rv_{cr} \qquad (5.36)$$

where B_0 is the average magnetic field. The diffusion is highly anisotropic with $K_{||} \gg K_\perp$.

If $K_{||}$ is a constant out to the edge of the modulating region, the cosmic ray density at the earth is

$$n(r_{se}, E) = n(\infty, E) \exp(-\eta/Rv_{cr}) \qquad (5.37)$$

where the magnetic fluctuation diffusion variable η is evaluated from Eq. 5.32. The long-term change in the cosmic ray flux results from a change in η, such that the ratios of the cosmic ray densities at two different times is

$$n_2/n_1 = \exp\left[(\eta_1 - \eta_2)/Rv_{cr}\right] \qquad (5.38)$$

Since the rigidy of alpha-particles is twice the rigidity of protons at the same velocity, the diffusion coefficient for protons should be twice as large as for alpha-particles. And indeed this is true for protons and alpha-particles measured on Imp 2 and 3.[61,62]

The observations near the earth only measure the dependence of the change in modulation on the velocity and magnetic rigidity. Measurements at different distances from the sun are required to obtain the magnitude of the modulation. The necessary cosmic ray gradient measurements were made on Mariner 4 by J. O'Gallagher and J. Simpson.[62] From December 1964 to September 1965 cosmic ray protons and alpha-particles were measured over a distance of 0.6 AU. In Fig. 5.27 the ratio of the cosmic ray density measured by Mariner 4 to that measured at the earth is plotted against the distance from the earth. The fractional change in cosmic ray density with distance at a

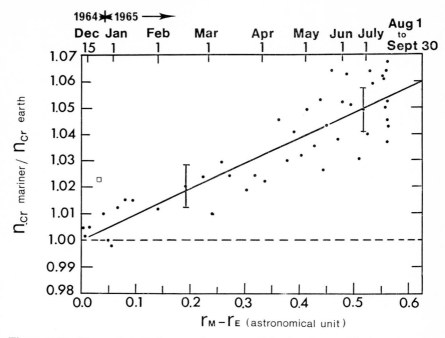

Figure 5.27. The ratio of the cosmic ray particle density at Mariner 4 to the density at the earth outside of the earth's magnetic cavity versus the distance from the earth. The fractional change in cosmic ray density with distance is $0.096 \pm 0.009)$/AU. From O'Gallagher and Simpson,[62] 1967.

rigidity of 6 BV is the slope of the line, $g = (0.096 \pm 0.009)$/AU. Substituting the measured gradient into Eq. 5.34 along with the solar wind velocity of 400 km/sec gives

$$K_{\parallel} \simeq 1.3 \times 10^{22} \text{ cm}^2/\text{sec} \qquad (5.39)$$

at a cosmic ray magnetic rigidity of 6 BV. If this measured value of the radial gradient of the cosmic rays extends out to several AU, the low energy protons and alpha-particles will be much more abundant outside the solar wind cavity than at the earth.

The authors also pointed out that the two separate measurements, the radial gradient of the cosmic ray density and the density measurements at different times of the solar cycle, show that the diffusion coefficient is separable into two functions. One is a function of rigidity and velocity only and the other a function of distance and time only

$$K(Rv_{\text{cr}}rt) = K_1 (Rv_{\text{cr}}) K_2 (rt) \qquad (5.40)$$

We now return to equation 5.32 to evaluate the distance $\mathscr{L}$ over which the cosmic ray modulation is effective. The various methods for determining $\mathscr{L}$ seem to be approaching agreement at about 5 AU at solar minimum.[63] In one method, the value of the cosmic ray density gradient is used to extrapolate outward from the sun until the magnetic field density is down to the value in interplanetary space. This gives $\mathscr{L} \simeq 5$ AU.

The diffusion coefficient $K_{||}$ is proportional to the rigidity and the rigidity equals

$$R = \frac{A}{Z} f(v). \tag{5.41}$$

The A is the particle mass, Z the charge and $f(v)$ is a function of the velocity. The helium and deuterium ratios, $He^3/(He^3 + He^4)$ and H^2/He^4, in cosmic rays are modulated differently in the fluctuating magnetic fields of interplanetary space because of the rigidity dependence on A and Z. From these ratios and the cosmic ray density gradient, $\mathscr{L}$ is estimated to be < 5 AU.

An early method to obtain $\mathscr{L}$ used the phase lag of the cosmic ray density relative to the time of the solar minimum. The cosmic ray maximum as measured by the neutron monitors on the surface of the earth followed the sunspot minimum by a time Δt. It was argued that the time delay is the time for the solar wind with velocity v_s to travel the distance $\mathscr{L}$ from the sun to the boundary of the solar wind cavity. So $\mathscr{L} = \Delta t \times v_s$. Early estimates of the phase lag were about one year and these estimates gave values of $\mathscr{L}$ as large as 100 AU. However it is now maintained that the line intensities at wavelengths of visible light better measure the changing solar wind characteristics than the sunspot number or solar radio emission.[63] The phase lag from these data is now reduced to 1 or 2 months and gives a value of $\mathscr{L}$ from 5 to 10 AU. All methods are now in reasonable agreement.

In summary, the constants of diffusion of cosmic ray particles in inter-planetary space are: (1) The density gradient $(dn/dr)/n$. It is 0.09/AU for energies of 6 BeV, 0.7 for energies of 100 to 420 MeV and 2.1 for energies of 20 to 100 MeV[64]. (2) The diffusion constant along the magnetic field lines $K_{||}$ is 1.3×10^{22} cm²/sec for a particle with Rv/c of 6 BV and 3.2×10^{21} cm²/sec for a particle of 1 BV. (3) The best value for the extent of the modulation region for the time of solar minimum around 1965 is 5 AU.

5.9 PROPAGATION OF LOW ENERGY SOLAR PROTONS AND ELECTRONS

It has been recently observed[65] that low energy solar protons and electrons—protons with energies greater than about 0.5 MeV and electrons with energies greater than about 40 keV—propagate through interplanetary space in rather wide streams, much wider than are expected from point flare sources on the sun. In one type of event, electrons arrive at the earth from 30 to 60 minutes after the maximum of the solar flare with a beam width of 30 deg. They come from solar flare sources that lie between 45 and 75 deg W. solar longitude. The electrons are highly anisotropic for times of 10 to 15 hours and have directions primarily along the magnetic field lines away from the sun.

In a second type of event the electrons originate in flares beneath the radio noise regions. These electron beams have much wider angular widths at the earth—about 90 deg. The electrons originate from E 28 to W 10 solar longitudes and can only reach the earth when a large type 1 noise region is present. It is not clear whether the wide beam is caused by electrons that are released over a large area of the sun or whether the electrons are produced at a point and then spread over a large area on the sun's surface before escaping into interplanetary space.

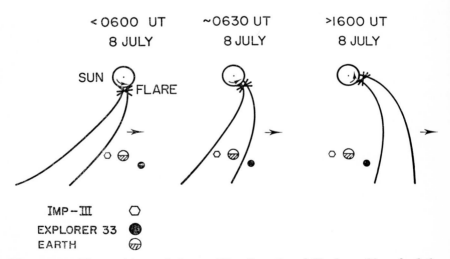

Figure 5.28. The positions of the satellites Imp 3 and Explorer 33 and of the earth relative to the solar beam of electrons and protons at three times on July 8, 1966. The diagram shows how the beam sweeps by the satellites and the earth. From Lin, Kahler and Roelof,[66] 1968.

Many of the solar flares also produce low energy protons. While the electron beams die away with a characteristic time of several hours, the protons undergo large changes in an hour. It is not certain whether the protons are produced and accelerated with the electrons.

Simultaneous electron and proton fluxes were measured on the three satellites Imp 3, OGO 3 and Explorer 33. By comparing measurements from these satellites, R. Lin, S. Kahler and E. Roelof[66] determined the angular extent of the particle beams from the sun. The positions of the satellites Imp 3 and Explorer 33 and the earth are shown in

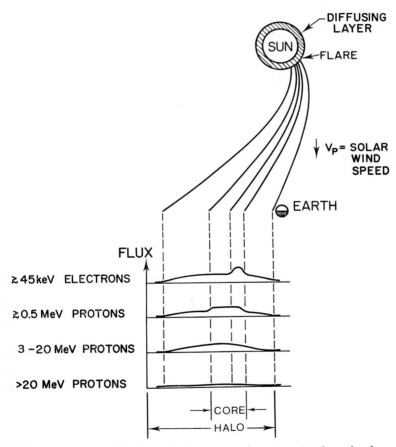

Figure 5.29. Space distributions of electrons and protons in the solar beam on July 8, 1966. Distributions are shown for electrons with energies > 45 keV and for protons with energies > 0.5 MeV, 3–20 MeV and > 20 MeV. The intense core and diffuse halo are indicated. From Lin, Kahler and Roelof,[66] 1968.

I

Fig. 5.28 for July 8, 1966. The particle beam is shown before, during and after its sweep across the earth.

The intense flux of electrons and protons in the central core of the beam and the less intense halo on the fringes are shown in Fig. 5.29. The authors[66] suggest that the particle flux distribution is the same as the distribution injected at the sun. Some of the particles are injected quickly while others are temporarily stored. The distribution travels outward along the curved magnetic field lines. The core is connected to the main particle production region in the flare and the halo to the particles that are temporarily stored and slowly leak out onto the interplanetary magnetic field lines.

Additional evidence for the wide beams comes from the Pioneer 6 and 7 space probes.[67] Proton beams with energies of 0.6 to 13 MeV sometimes are injected with angular widths as great as 180 deg. The protons start at 60 to 70 deg east of the sun–earth line and stop at 100 to 130 deg west. These regions rotate with the sun. Superimposed on the low energy protons are the occasional beams of higher energy protons of tens of MeV from solar flares. The flare protons are injected over solar longitudes of greater than 60 deg.

C. Fan, M. Pick, R. Pyle, J. Simpson and A. Smith suggest[67] that the magnetic field lines originate in the active proton source and spread through 100 to 180 deg in the corona. The magnetic field lines are then carried into interplanetary space by the solar wind. The protons follow these spread-out magnetic field lines and appear to originate from a region about 180 deg wide. The model of the field lines is given in Fig. 5.30. The region inside of the dashed line gives the spread-out magnetic field. In the region outside the dashed line the magnetic field lines follow the familiar Archimedes spiral. The position of the leading edge of the proton beam, the two flare events b and f and the cut-off are also shown.

An explanation for the wide beams of particles was given by J. Jokipii and Eugene Parker.[68] Random fluctuations in the magnetic fields close to the surface of the sun lead to a rapid diffusion of energetic particles across the average magnetic field. While the particles move along the direction of the field z with velocity v, they also diffuse perpendicular to the average magnetic field, x, with an average diffusion coefficient

$$K_\perp = \frac{\overline{(\varDelta x)^2}}{\varDelta t} = v_\parallel \frac{\overline{(\varDelta x)^2}}{\varDelta z} \qquad (5.42)$$

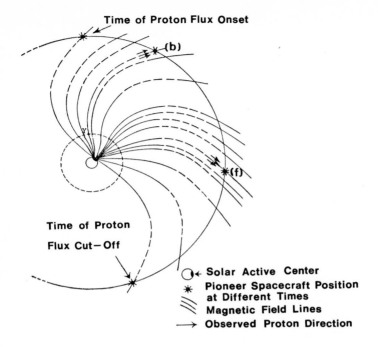

Figure 5.30. Model for the connection of the active region on the sun to the interplanetary magnetic field. The dashed circle indicates the height at which the Archimedes spiral magnetic field begins. The pioneer spacecraft symbols indicate the positions where the proton flux begins, events b and f, and where the proton flux stops. The dashed part of the magnetic field curve indicates a shortening of the distance scale in that region. The arrows indicate the direction of the protons. From Fan, Pick, Pyle, Simpson and Smith,[67] 1968.

The random walk of the magnetic field lines is related to the magnetic field power distribution at zero frequency $P_{xx}(0)$ gauss²-sec by

$$\frac{\overline{(\Delta x)^2}}{\Delta z} = \frac{v_s}{B_0^2} P_{xx}(0). \tag{5.43}$$

B_0 is the average magnetic field. Using $P_{xx}(0) = 1.25 \times 10^3 \, B_0^2$, they find

$$\frac{\overline{\Delta x^2}}{\Delta z} = 5 \times 10^{10} \text{ cm} \tag{5.44}$$

At the orbit of the earth where $\Delta z = 1$ AU, the root mean square diffusion width is

$$\sqrt{\overline{(\Delta x)^2}} = 1.5 \times 10^{12} \text{ cm} = 0.1 \text{ AU} \tag{5.45}$$

The random walk is very likely produced by the granule and super-granule motions of the photosphere of the sun. For a description see section 4.4 and a review of supergranulation by Leighton.[69] The super-granules have a width L_{sg} of about 1.5×10^4 km and a lifetime Δt_{sg} of about 3×10^4 sec. At the photosphere these values give a diffusion coefficient perpendicular to the magnetic field of

$$K_\perp = \frac{L_{sg}^2}{\Delta t_{sg}} = 0.75 \times 10^{14} \text{ cm}^2/\text{sec} \qquad (5.46)$$

It takes the solar wind with the velocity of 400 km/sec, 4×10^5 sec to reach the earth. In this time N_{sg} supergranulations will be formed and the feet of the magnetic field lines will random walk a distance of

$$\sqrt{N_{sg}} \, L_{sg} = \sqrt{4 \times 10^5/3 \times 10^4} \times 1.5 \times 10^4 \text{ km} = 5.5 \times 10^4 \text{ km}.$$

The angular displacement at the sun is 0.08 rad. The displacement at the earth is then $0.08 \times 1.5 \times 10^{13}$ cm $= 1.2 \times 10^{12}$ cm in good agreement with the value calculated above from the magnetic field disturbances observed on Mariners 2 and 4.

The supergranulation length L_{sg} is magnified at the earth to $220 L_{sg}$ or 3×10^{11} cm. This is just the dimension of a tube of force observed at the earth by the cosmic ray detector on Pioneer 6.[70] The wide angle beams up to 180 deg observed by the Chicago proton detector[67] are explained by this perpendicular random walk diffusion. The calculated full width of the gaussian curve at $1/e$ of the maximum is 33 deg. The gaussian roughly reproduces the beam fluxes for the measured full width of 180 deg.

5.10 SOLAR WIND INTERACTIONS WITH THE MOON

What happens to the solar wind and the interplanetary magnetic field near large objects like the earth or the moon? In chapter 6 we will review the formation of the magnetic cavity in the vicinity of the earth. We will discuss the spherical nose in the direction of the sun, the long cylindrical tail pointing away from the sun and the bow shock that is formed as the solar wind runs into the magnetic cavity. But in this section we will limit our discussion to the interaction of the moon with the solar wind and the interplanetary magnetic fields.

The moon has little or no permanent magnetic field. Its upper limit at an altitude of 0.5 to 1 moon radius is $2\,\gamma$. A hypothetical lunar magnetic dipole in the direction of the moon's rotation axis is less than 7×10^{-6} of that of the earth. That limit was obtained by the Ames magnetometer experiment[71] on Explorer 35.

Suppose the moon were a perfect conductor, even better than a copper sphere. The interplanetary magnetic field would not be able to penetrate because of opposing magnetic fields set up by the induced currents that exactly cancel the penetrating magnetic field. The magnetic field lines would then pile up on the sun side of the moon. The conditions near the moon would be similar to those at the earth. There would be a forward bow shock, a magnetic cavity and a transition region in between of about 30 gauss. Since the solar wind velocity is larger than the velocity of the transmission of information in the region, a shock wave upstream from the moon would be formed.

In the opposite case where the moon is a perfect insulator, a non-conducting sphere, the magnetic field lines would pass through the moon unimpeded. The solar wind plasma would be stopped by the moon and a vacuum without plasma would be formed in the geometrical shadow of the moon. In this case no forward bow shock would be formed. A similar situation would occur for a conducting sphere with an insulating layer on the surface.

The moon appears to be more like an insulator. The plasma experiment of E. Lyon, H. Bridge and J. Binsack on Explorer 35[72] measured the flux of protons of 50 to 540 eV as it passed close to the moon. On the sun side the particle fluxes were the same as in interplanetary space. There was no evidence of a change in flux or a disordered motion of the particles that is characteristic of passing through the bow shock. However, when the experiment passed into the geometric shadow behind the moon the flux of particles fell rapidly below the threshold of the detector. The experiment discovered the conical vacuum region without particles behind the moon.

On passing through the shadow, the NASA-Ames magnetometer experiment[71] found a short increase of about 1 γ, followed by a decrease of 2 or 3 γ then an increase to a plateau of about 2 γ above the interplanetary field. This plateau lasted until the spacecraft left the shadow cone where the pattern repeated in a symmetrical way. The same type of pattern was also observed by the NASA-GSFC magnetic field experiment[73] on Explorer 33.

The magnetic field changes are explained by the solar wind shadow cast by the moon. As the magnetic field lines outside the shadow try

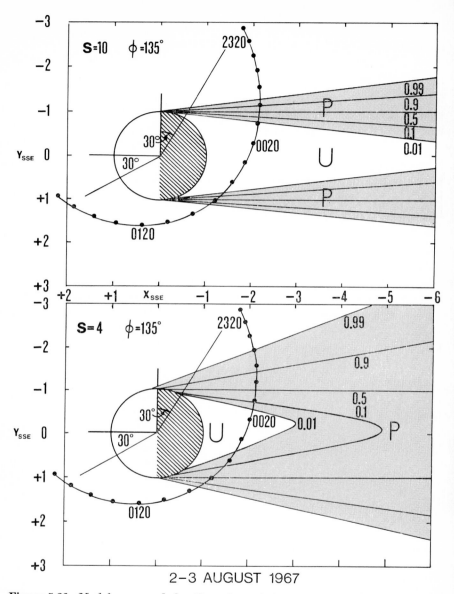

Figure 5.31. Model proposed for the solar wind density decrease behind the moon. The interplanetary magnetic field is taken to lie in the ecliptic plane at an angle of 55 deg to the sun–earth line. The numbers 0.99 to 0.01 on the contours are the fractions of the undisturbed plasma flux behind the moon. The shaded regions of the penumbra P starts with the contour 0.99 and ends with the contour 0.01. The open region of the umbra U includes all the region for the fraction < 0.01. S is the ratio of the solar wind velocity to the thermal velocity parallel to the magnetic field direction. The orbit of Explorer 35 is projected into the ecliptic plane behind the moon. From Ness, Behannon, Taylor and Whang,[73] 1968.

to move into the shadow, currents are set up to oppose that change. The magnetic field at the boundary decreases. As the boundary moves into the shadow, the volume available to magnetic field lines in the shadow decreases. So the magnetic field in the shadow increases to the plateau value observed. The slight positive increase measured on first crossing the boundary is unexplained.

A sketch for one model of the shadow region is given in Fig. 5.31.[73] Two views in the ecliptic plane are shown. For the upper view a large ratio of the solar wind velocity to the thermal velocity parallel to the magnetic field S was used and for the lower view a smaller ratio was selected. The trajectory of the spacecraft is shown with the beaded line. The umbra and penumbra are indicated with the symbols U and P. The penumbral region is shaded. The sun is at the left. It appears that the exact variations of the magnetic field depend critically on the solar wind velocity, density, temperature and magnetic field orientation. Repeated measurements from many orbits around the moon give no evidence for a forward bow shock or a standing shock behind the moon. However, it is still possible that a trailing shock exists and will be seen by experiments with orbits that pass through the shadow at greater distances from the moon.

Great improvements in our knowledge of interplanetary space have been made with experiments on space probes. Observations have been made that verify some ideas that are more than 70 years old. The observations have often verified the fragmentary or secondary information from measurements on the surface of the earth. Frequently the space experiments of the last 10 years have made new and important discoveries that were never anticipated. In the next years, particularly during the oncoming solar cycle, we are sure that more new and unpredicted discoveries will be made. Discoveries that will answer many of our questions about the origin and history of the solar system. Discoveries that are beyond our present imagination.

REFERENCES

1. Fitzgerald, George F., "Sunspots and magnetic storms," *The Electrician*, November 11, p. 48 (1892).
2. Fitzgerald, George F., 'Sunspots, magnetic storms, comets tails, atmospheric electricity, and aurorae," *The Electrician*, December 14, p. 287 (1900).
3. Birkeland, Kr., "Sur les rayons cathodiques sons l'action de forces magnétiques intenses," Arch. Sci. Phys. Naturelles, **1**, 497–512 (1896).

4. Birkeland, Kr., "The Norwegian aurora polaris expedition 1902–3," Vol. 1, *On the Cause of Magnetic Storms and the Origin of Terrestrial Magnetism,* first section, (1908) and second section (1913), H. Aschehoug and Co., Christiania (1913).

5. Dessler, Alexander J., "Solar wind and interplanetary magnetic field," *Reviews of Geophysics* **5**, 1–41 (1967).

6. Chapman, S. and Bartels, J., *"Geomagnetism,"* Oxford Press, London (1940).

7. Chapman, S. and Ferraro, V. C. A., "A new theory of magnetic storms, Part I—The initial phase, and Part II—The main phase," *Terrest. Magnetism and Atmospheric Electricity* **36**, 77–97, 171–186 (1931); **37**, 147–156, 421–429 (1932); **38**, 79–96 (1933).

8. Forbush, Scott E., "On world-wide changes in cosmic-ray intensity," *Phys. Rev.* **54**, 975 (1938).

9. Morrison, Philip, "Solar origin of cosmic ray time variations," *Phys. Rev.* **101**, 1397–1404 (1954).

10. Behr, A. and Siedentopf, H., "Untersuchen über Zodiakallicht und Gegenschein nach lichtelektrischen Messungen auf dem Jungfraujoch," *Zeitschr. f. Astrophys.* **32**, 19–50 (1953).

11. Blackwell, D. E. and Ingham, M. F., *Monthly Not. Roy. Astr. Soc.* **122**, 113 (1961).

12. Lüst, R., "Interplanetary plasma," *Space Science Rev.* **1**, 522–552 (1963).

13. Biermann, L., "Kometenschweife und Solare Korpuskularstrahlung," *Z. Astrophys.* **29**, 274–286 (1951).

14. Alfvén, H., "On the theory of comet tails," *Tellus,* **9**, 92–96 (1957).

15. Biermann, L. and Lüst, Rh., *"Comets: structure and dynamics of tails in the solar system, IV,"* edited by B. Middlehurst and G. P. Kuiper, Chapter 18, University of Chicago Press, Chicago (1963).

16. Chapman, S., "Interplanetary space and the earth's outermost atmosphere," *Proc. Roy. Soc. London* **A253**, 462–481 (1959).

17. Parker, E. N., "The solar wind," *Scientific American,* April, p. 66–76 (1964).

18. Parker, E. N., "Dynamics of the interplanetary gas and magnetic fields," *Astrophys. J.* **128**, 664–676 (1958).

19. Parker, E. N., "Dynamical theory of the solar wind," *Space Science Reviews* **4**, 666–707 (1965).

20. Clauser, Francis H., "The aerodynamics of mass loss and mass gain of stars," *Johns Hopkins University Lab. Rept.,* AFOSR TN 60–1386, Nov. (1960).

21. Alfvén, H., "Granulation, magneto-hydrodynamic waves, and the heating of the solar corona," *Monthly Notices Roy. Astron. Soc.* **107**, 211–219 (1947).

22. Gringauz, K. I., Bezrukikh, V. V., Ozerov, V. D. and Kybchinskii, R. E., "Study of the interplanetary ionized gas, high energy electrons, and solar corpuscular radiation by means of three electrode traps for charged particles on the second cosmic rocket," *Soviet Phys. Doklady* **5**, 361–364 (1964); or *Dokl. Akad. Nauk. SSSR* **131**, 1301–1304 (1960).

23. Gringauz, K. I., "Some results of experiments in interplanetary space by means of charged particle traps on soviet space probes," *Space Res.* **2**, 539–553 (1961); or *Iskusstoenny, Sputniki Zeonti* **12**, 119–132 (1962).

24. Bonetti, A., Bridge, H. S., Lazarus, A. J., Rossi, B. and Scherb, F., "Explorer 10 plasma measurements," *J. Geophys. Res.* **68**, 4017–4063 (1963).

25. Neugebauer, M. and Snyder, C. W., "The mission of Mariner 2: preliminary observations, solar plasma experiment," *Science* **138**, 1095–1097 (1962).

26. Neugebauer, Marcia and Snyder, Conway W., "Mariner 2 observations of the solar wind 1. average properties," *J. Geophys. Res.* **71**, 4469–4484 (1966).

27. Snyder, C. W., Neugebauer, M. and Rao, U. R., "The solar wind velocity and its correlation with cosmic-ray variations and with solar and geomagnetic activity," *J. Geophys. Res.* **68**, 6361–6370 (1963).

28. Hundhausen, A. J., Asbridge, J. R., Bame, S. J., Gilbert, H. E. and Strong, I. B., "Vela 3 satellite observations of solar wind ions: a preliminary report," *J. Geophys. Res.* **72**, 87–100 (1967).

29. Neugebauer, M. and Snyder, C. W., "Mariner 2 measurements of the solar wind," in the *Solar Wind*, edited by R. J. Mackin and M. Neugebauer, p. 3–21, Pergamon Press, New York (1966).

30. Biswas, S. and Fichtel, C. E., "Nuclear composition and rigidity spectra of solar cosmic rays," *Astrophys. J.* **139**, 941–950 (1964).

31. Chapman, S., "Solar streams of corpuscles: their geometry, absorption of light, and penetration," *Royal Astronomical Soc. Monthly Notices* **89**, 456–469 (1929).

32. Ahluwalia, H. S. and Dessler, A. J., "Diurnal variation of cosmic radiation intensity produced by a solar wind," *Planetary Space Sci.* **9**, 195–210 (1962).

33. Coleman, P. J., Jr., Davis, Leverett, Jr. and Sonett, C. P., "Steady component of the interplanetary magnetic field: Pioneer V," *Phys. Rev. Letters* **5**, 43–46 (1960).

34. Coleman, P. J., Jr., Sonett, C. P., and Davis, L., Jr., "On the interplanetary magnetic storm: Pioneer V," *J. Geophys. Res.* **66**, 2043–2046 (1961).

35. Coleman, P. J., Jr., Davis, Leverett, Jr., Smith, E. J. and Sonett, C. P., "Interplanetary magnetic fields," *Science* **138**, 1099–1100 (1962).

36. Heppner, J. P., Ness, N. F., Scearce, C. S. and Skillman, T. L., "Explorer 10 magnetic field measurements," *J. Geophys. Res.* **68**, 1–46 (1963).

37. Ness, Norman F., Scearce, Clell S. and Seek, Joseph B., "Initial results of the Imp 1 magnetic field experiment," *J. Geophys. Res.* **69**, 3531–3569 (1964).

38. Ness, Norman F. and Wilcox, John M., "Sector structure of the quiet interplanetary magnetic field," *Science* **148**, 1592–1594 (1965).

39. McCracken, K. G., "The cosmic-ray effect 3. deductions regarding the interplanetary magnetic field," *J. Geophys. Res.* **67**, 447–458 (1962).

40. Wilcox, John M., "Solar and interplanetary magnetic fields," *Science* **152**, 161–166 (1966).

41. Fairfield, D. H. and Ness, N. F., "Magnetic field measurements with the Imp 2 satellite," *J. Geophys. Res.* **72**, 2379–2402 (1967).

42. Fan, C. Y., Gloeckler, G. and Simpson, J. A., "Protons and helium nuclei within interplanetary magnetic regions which co-rotate with the sun," *Proc. Int. Conf. Cosmic Rays*, p. 109–111 (1965).

43. Wilcox, John M. and Ness, Norman F., "Quasi-stationary corotating structure in the interplanetary medium," *J. Geophys. Res.* **70**, 5793–5805 (1965).

44. Bryant, D. A., Cline, T. L., Desai, U. D. and McDonald, F. B., "New evidence for long-lived solar streams in interplanetary space," *Phys. Rev. Letters* **11**, 144–146 (1963).

45. Bartels, J., *Terrest. Magnetism Atm. Elec.* **37**, 48 (1932).

46. Bryant, D. A., Cline, T. L., Desai, U. D. and McDonald, F. B., "Continual acceleration of solar protons in the MeV range," *Phys. Rev. Letters* **14**, 481–484 (1965).

47. O'Gallagher, J. J. and Simpson, J. A., "Anisotropic propagation of solar protons deduced from simultaneous observations by earth satellites and the Mariner 4 space probe," *Phys. Rev. Letters* **16**, 1212–1217 (1966).

48. McCracken, K. G. and Ness, N. F., "The collimation of cosmic rays by the interplanetary magnetic field," *J. Geophys. Res.* **71**, 3315–3318 (1966).

49. McCracken, K. G., Rao, U. R. and Bukata, R. P., 'Recurrent Forbush decreases associated with M-region magnetic storms," *Phys. Rev. Letters* **17**, 929–932 (1966).

50. Wolfe, J. H., Silva, R. W., McKibbin, D. D. and Mason, R. H., "The compositional, anisotropic, and nonradial flow characteristics of the solar wind," *J. Geophys. Res.* **71**, 3329–3335 (1966).

51. Sonett, C. P., Colburn, D. S., Davis, L. Jr., Smith, E. J. and Coleman, P. J. Jr., "Evidence for a collision-free magnetohydrodynamic shock in interplanetary space," *Phys. Rev. Letters* **13**, 153–156 (1962).

52. Gosling, J. T., Asbridge, J. R., Bame, S. J., Hundhausen, A. J. and Strong, I. B., "Measurements of the interplanetary solar wind during the large geomagnetic storm of April 17–18, 1965," *J. Geophys. Res.* **72**, 1813–1832 (1967).

53. Gosling, J. T., Asbridge, J. R., Bame, S. J., Hundhausen, A. J. and Strong, I. B., "Satellite observations of interplanetary shock waves," *J. Geophys. Res.* **73**, 43–50, 1968.

54. Coleman, Paul J., Jr., "Variations in the interplanetary magnetic field: Mariner 2, 1. observed properties," *J. Geophys. Res.* **71**, 5509–5531 (1966).

55. Siscoe, G. L., Davis, L., Jr., Coleman, P. J., Jr., Smith, E. J. and Jones, D. E. "Power spectra and discontinuities of the interplanetary magnetic field Mariner 4," *J. Geophys. Res.* **73**, 61–82 (1968).

56. Colburn, D. S. and Sonett, C. P., "Discontinuities in the solar wind," *Space Sci. Rev.* **5**, 439–506 (1966).

57. Parker, E. N., *Interplanetary Dynamical Processes*, Interscience Publishers, New York (1963).

58. Axford, W. I., "The modulation of galactic cosmic rays in the interplanetary medium," *Planetary and Space Science* **13**, 115–130 (1965).

59. Gloeckler, G. and Jokipii, J. R., "Low-energy cosmic-ray modulation related to observed interplanetary magnetic field irregularities," *Phys. Rev. Letters* **17**, 203–207 (1966).

60. Jokipii, J. R., "Cosmic-ray propagation. I. charged particles in a random magnetic field," *Ap. J.* **146**, 480–487 (1966).

61. Fan, C. Y., Gloeckler, G., Hsieh, K. C. and Simpson, J. A., "Isotopic abundances and energy spectra of He^3 and He^4 above 40 MeV per nucleon from the galaxy," *Phys. Rev. Letters* **16**,, 813 (1966).

62. O'Gallagher, J. J. and Simpson, J. A., "The heliocentric intensity gradients of cosmic-ray protons and helium during minimum solar modulation," *Ap. J.* **147**, 819–827 (1967).

63. Simpson, J. A. and Wang, J. R., "Dimension of the cosmic ray modulation region," *Ap. J.* **149**, L73–L78 (1967).

64. O'Gallagher, J. J., "Cosmic-ray radial density gradient and its rigidity dependence observed at solar minimum on Mariner IV," *Ap. J.* **150**, 675–698 (1967).

65. Lin, R. P. and Anderson, K. A., "Electrons > 40 keV and protons > 500 keV of solar origin," *Solar Physics* **1**, 446–464 (1967).

66. Lin, R. P., Kahler, S. W. and Roelof, E. C., "Solar flare injection and propagation of low energy protons and electrons in the event of 7–9 July, 1966," to be published in *Solar Physics* (1968).

67. Fan, C. Y., Pick, M., Pyle, R., Simpson, J. A. and Smith, D. R., "Protons associated with centers of solar activity and their propagation in interplanetary magnetic field regions corotating with the sun," *J. Geophys. Res.* **73**, 1555–1582 (1968).
68. Jokipii, J. R. and Parker, E. N., "Random walk of magnetic field lines of force in astrophysics," *Phys. Rev. Letters* **21,**, 44–47 (1968).
69. Leighton, R. B., "The solar granulation," from *Annual Review of Astronomy and Astrophysics*, Vol. 1, 19–40 (1963). Published by Annual Reviews, Inc., Palo Alto, California.
70. Bartley, W. C., Bukata, R. P., McCracken, K. G. and Rao, U.R., "Anisotropic cosmic radiation fluxes of solar origin," *J. Geophys. Res.* **71**, 3297–3304 (1966).
71. Colburn, D. S., Currie, R. G., Mihalov, J. D. and Sonett, C. P., "Diamagnetic solar-wind cavity discovered behind moon," *Science* **158**, 1040–1042 (1967).
72. Lyon, E. F., Bridge, H. S. and Binsack, J. H., "Explorer 35 plasma measurements in the vicinity of the moon," *J. Geophys. Res.* **72**, 6113–6117 (1967).
73. Ness, N. F., Behannon, K. W., Taylor, H. E. and Whang, Y. C., "Perturbations of the interplanetary magnetic field by the lunar wake," *J. Geophys. Res.* **73**, 3421–3440 (1968).

CHAPTER 6

The Earth's Magnetic Cavity (Magnetosphere)

6.1 INTRODUCTION

We are all familiar with the earth's magnetic field. The Chinese used the magnetic compass for land journeys as early as 1,000 B.C.[1] But it wasn't until the 14th century that maps of the Mediterranean coast line and of Western Europe were made with compass bearings. The compass was used by Columbus on his first voyage to America in 1492 and his sailors were terrified when they discovered that it no longer pointed true north. They had discovered magnetic declination.

The real impetus to the study of terrestrial magnetism came from Gilbert in 1600 with his publication of "De Magnete." He proposed that the earth was an enormous spherical lodestone that had magnetic poles and a magnetic equator just like the model "terrella" or little earth that he had constructed. He pointed out that the earth itself simulates a spherical magnet. However, the theory was discarded when analysis of surface rocks found comparatively little magnetite or other magnetic material.

A mathematical description of the earth's magnetic field was formulated by the great mathematicians Laplace, Poisson, and Gauss. Gauss introduced the potential theory and the spherical harmonic analysis that form the basis for the description of the magnetic field today. In 1940, Chapman and Bartels[1] in their classic book, "Geomagnetism," described their ideas of terrestrial magnetism and magnetic storms and reviewed the subject to that date.

Since Gilbert's time the most able men of science have attacked the problem of the origin of terrestrial magnetism. But it still remains one of the unsolved scientific mysteries of today.

6.2 MAGNETIC FIELD MEASUREMENTS

The first measurements of the earth's outer magnetic field were made by Charles Sonett, D. Judge and J. Kelso[2] on Pioneer I. On a trajectory near the sun–earth line, the magnetic field was found to be close to an undistorted dipole from 3 to $7R_e$. However, at greater distances it was larger than expected. At 12 to $14R_e$ the field fluctuated wildly and then decreased abruptly. Two years later, measurements[3] on Pioneer V on a 3:00 p.m. trajectory gave similar results with fluctuating fields beyond $10R_e$ and a decreased field from 15 to 20 R_e. The trajectories of these and subsequent satellites are shown in Fig. 6.1[4]

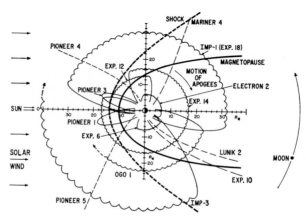

Figure 6.1. The trajectory of satellites and space probes, projected on the ecliptic plane, taken from Ness,[4] 1966. Positions of the geomagnetic field boundary, the shock front, and the moon are shown. The solar wind direction is incident from the left.

projected onto the ecliptic plane. The ripples in the orbits show the progression of the apogees of the satellites in a clockwise direction.

The first real evidence for the non-dipole nature of the earth's magnetic field and for a comet-like tail in the direction away from the sun came from Explorer 10. Heppner and his associates[5] measured the magnetic field magnitude and direction in a trajectory along the 9:00 p.m. direction. The field lines appeared to be pulled away from the earth. At $21R_e$ the magnetic field dropped abruptly from 20 to 10 γ (1 $\gamma = 10^{-5}$ gauss) and the direction of the field changed by nearly 180 deg. They discovered the boundary between the earth's magnetic field tail and interplanetary space. Additional evidence came from Explorers 12[6] and 14[7] which threaded in and out through the boundary. These results are described and summarized in the review by Cahill.[8]

Beautiful and convincing measurements of the magnetic field cavity boundary, and of fields in the tail and in interplanetary space were carried out by Norman Ness, C. Scearce, and J. Seek[9] on Imp 1 (Explorer 18). As can be seen from Fig. 6.1, Imp 1 was launched at 30 deg to the earth–sun line. Its trajectory precessed 180 deg while crossing the magnetic cavity boundary many times. An example of the magnetometer data from orbit 1 of Imp 1[4] taken while crossing the boundary is shown in Fig. 6.2. The magnitude of the magnetic field $\bar{F}$,

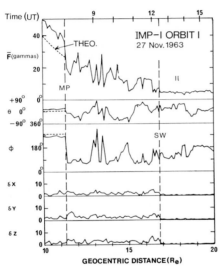

Figure 6.2. Magnetic field measurements on the outbound part of orbit one of the Imp 1 satellite on November 27, 1963, taken from Ness,[4] 1966. The magnitude of the magnetic field $\bar{F}$, the angle of the magnetic field with respect to the sun–earth line in the meridian plane θ, the angle with respect to the sun–earth line in the ecliptic plane ϕ, and the root mean square deviations of the three rectangular magnetic field values from their mean values are plotted versus distance from the earth. The magnetosphere boundaries at $11.3R_e$ and the termination of the rapid fluctuations of the magnetic field at $16.8R_e$ at the shock front are visible.

the angle of the magnetic field with respect to the sun–earth line in the noon–midnight meridian plane θ, and the angle with respect to the sun–earth line in the ecliptic plane of the earth ϕ, are plotted against time. The magnetic cavity boundary is clearly visible at $11.3R_e$. The end of the region with rapid fluctuations at $16.8R_e$ is the shock front boundary which separates interplanetary space from the magneto-sheath, the region between the magnetic cavity and interplanetary space.

Comparisons of the earth's magnetic field as known before and after 1961 are shown in Fig. 6.3 and 6.4. Figure 6.3 gives the usual dipole field that falls off inversely as the third power of the distance from the center of the earth and Fig. 6.4 shows a cross-section through the magnetic cavity. The boundary of the magnetic cavity is compressed to $10R_e$ in the sun–earth direction, enlarges to $14R_e$ at the poles and forms a long tail away from the sun. The tail has a radius of about $20R_e$ and extends for several earth–moon distances in the backward direction. Inside the magnetic cavity tail the magnetic field is 20 to 30 γ and the magnetic field energy density is larger than the particle kinetic energy density. Outside, the plasma energy density is larger.

A shock front is formed by the impact of the solar wind on the earth's magnetic field. This continuous stream of particles from the sun had been predicted in 1951 by Ludwig Biermann.[11] He suggested that

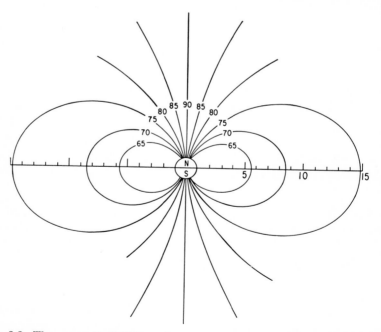

Figure 6.3. The magnetic field lines from a pure magnetic dipole. The magnetic field falls off as $1/R_e^3$. Angles of magnetic latitude are labeled. The horizontal scale is distance in earth radii.

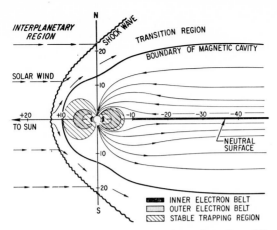

Figure 6.4. Section through the magnetic cavity taken from Ness,[10] 1965. The shock wave is shown as a wiggly line and the boundary of the magnetic cavity as a solid line with the transition region between. The very heavy line in the direction away from the sun indicates the neutral surface which separates the magnetic field in the northern half-cylinder of the magnetic tail from that of the southern half. The field lines reverse in crossing the neutral surface. The stable trapping region is indicated with slanted lines, the outer electron belt with dots, and the inner electron belt with a black area.

large particle fluxes from the sun would explain the observations that comet tails point away from the sun. In 1958 Eugene Parker[12] proposed that protons and electrons are continuously blown out from the sun by the hydrodynamic expansion of the solar corona. He named this particle flux the "solar wind." (See section 5.2.)

In 1960 the solar wind was detected experimentally by the Russian physicist K. Gringauz and his colleagues[13] and was verified in 1962 by the MIT Faraday Cup measurements on Explorer 10.[14] A good correlation between K_p, an indicator of magnetic activity, and the plasma velocity was found by Conway Snyder, Marcia Neugebauer, and V. Rao[15] with an electrostatic spectrometer on Mariner 2. The plasma velocity is

$$v_s \,(\text{km/sec}) = 8.44 \sum K_p + 330. \tag{6.1}$$

They found a further strong 27 day correlation which was associated with the period of rotation of the sun.

The shock front observed by the magnetic field measurements on Imp 1 expands from $14R_e$ in the sun–earth direction to about $20R_e$ at the poles. The solar wind is thermalized behind the shock front and compresses the magnetic field into a hemisphere on the sunlit side. Between the shock wave and the boundary of the magnetic cavity the magnetic fields fluctuate rapidly. This region of turbulence, the magnetosheath, is $4R_e$ thick in the sun–earth direction and $7R_e$ thick at the poles. The solar wind peels back the high latitude magnetic field lines near the poles to form the long tail away from the sun.

From the Imp 1 measurements, Ness[10] discovered a thin neutral sheet in the tail that was only 600 km thick. The neutral sheet separates the magnetic field that points toward the earth in the northern part of the tail from the field that points away from the earth in the south. The data from a typical orbit through the tail is shown in Fig. 6.5. The

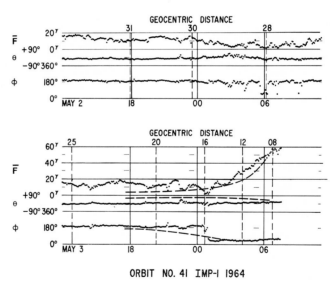

ORBIT NO. 41 IMP-1 1964

Figure 6.5. Magnetic field measurements from the inbound orbit 41 of Imp 1, May 2–4, 1964, taken from Ness,[10] 1965. The ecliptic coordinates $\bar{F}$, θ, and ϕ are plotted versus time. The magnetic field is directed away from the earth, $\phi = 180$ deg, to $16R_e$. There the neutral sheet is penetrated, the field reverses and then points toward the earth.

values of $\bar{F}$, θ, and ϕ are plotted against time. The spacecraft travels through the tail region near the magnetic field equator as it moves toward the earth. When it enters the neutral sheet the magnetic field approaches zero. On passing through the neutral sheet from south to north, ϕ abruptly reverses direction from 180 to 0 deg.

6.3 CHARGED PARTICLES

The boundaries of the different magnetic field regions have also been defined by charged particle detectors. Extensive measurements[16,17] near the magnetic field boundaries showed that the trapped particles are constrained to $10R_e$ on the sunlit side of the earth and to about $8R_e$ away from the sun. John Freeman,[17] from measurements on Explorer 12, reported that electrons with energy $E > 1.6$ MeV are peaked at L values of 4 to 5, that electrons with $E > 40$ keV extend out to $10R_e$ in the sun direction and that electrons from 200 eV to 40 keV reach out farther in both directions. These low energy electrons were earlier studied by Gringauz, et al.[13,18] with Faraday Cups on Luniks 2 and 1.

Particle experiments[19] were also carried out on Imp 1 with two Geiger tubes; one detected protons plus electrons directly and the other scattered electrons, only. Additional data and analysis[20] showed that the particles were grouped in time and that the build-up of the pulses took a few minutes but that the decay times were minutes to hours. T. Murayama[21] correlated fluxes of electrons with different tail variables and demonstrated convincingly that the electron fluxes decreased with distance from the neutral sheet. This is strong evidence that the electron sheet is the source of the tail electron pulses.

Extensive measurements were made by the Los Alamos Groups[22,23] on the Vela satellites. Six of these satellites at 16 to $18R_e$ steadily passed in and out of the magnetic cavity gathering boundary and tail information. The electrons were found clustered about the geomagnetic equatorial plane—near the neutral sheet. More electrons were seen on the dawn than on the dusk side of the magnetic cavity. The electron energy distributions were very steep; typically the fluxes changed by a factor of 1,000 when the electron energies changed from 50 to 200 keV.

The Vela electrostatic analyzer experiment[23] measured the thickness of the plasma sheet to be $6R_e$ in the tail at a distance of $17R_e$ from the earth. The electron energy distributions are nearly thermal and the particle energy densities are comparable to the magnetic field energy densities. The omnidirectional fluxes are usually between 10^8 and 10^9 electrons/cm²-sec. Ordinarily the high energy electrons with $E > 40$ keV are found at the same positions in space as the 0.35 to 20 keV electrons.

A three-dimensional drawing of the earth's magnetic cavity is given in Fig. 6.6. It shows the solar wind incident on the left, the shock front, the magnetosheath, and the earth's spherical magnetic cavity on the sunlit side. The maximum distance of the trapped electrons and protons from the earth in the sun's direction is $10R_e$, and in the direction

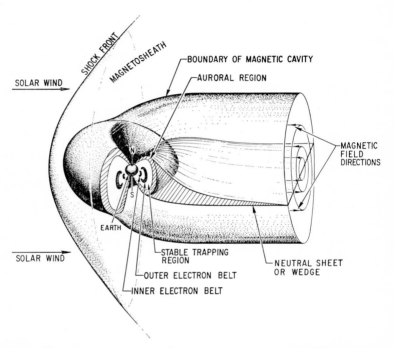

Figure 6.6. The magnetic cavity. It is spherical on the sunlit side and has a long tail $20R_e$, in radius, opposite the sun. A section shows the earth, the inner and outer electron radiation belts, and the region of stable trapping in the sun and anti-sun directions. The neutral sheet or wedge separates the upper half-cylinder where the field points toward the earth from the lower half-cylinder where it points away from the earth. The neutral wedge connects to the auroral region. The shock front between the solar wind and the magnetosheath is also shown.

away from the sun is $8R_e$. The earth with its North and South Poles and the inner and outer electron radiation belts are included in the cutaway. The cylindrical tail is shown with a radius of $20R_e$. The magnetic field points away from the earth below the neutral surface and toward the earth above.

6.4 MAGNETIC STORMS

Magnetic fields at the earth's surface vary as much as a few hundred gammas. These variations are called "magnetic storms." In the past these have been attributed to currents induced in the earth or the atmosphere, or to ring currents around the earth. After the discovery of electrons and protons in the 1890's, it was soon suggested that the sun was the source of the particles that caused the currents. Fitzgerald,[24] of the Lorentz–Fitzgerald contraction, and Lodge[25] suggested that magnetic storms were due to " 'a torrent or flying cloud of charged atoms or ions'; that auroras were caused by 'the cathode ray constituents . . . as they graze past the polar regions'; and that 'there seems to be some evidence from auroras and magnetic storms that the earth has a minute tail like that of a comet directed away from the sun.' "[26]

In 1933 Chapman and Ferraro[27] proposed a model to explain the magnetic field changes in magnetic storms. Particles evaporate from the sun, stream through interplanetary space, and are temporarily stored in the earth's magnetic field where they produce the currents that lower the earth's magnetic field. This large reduction in the magnetic field is called the main phase of the magnetic storm. After

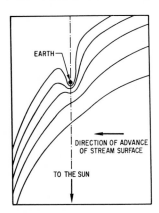

SUCCESSIVE EQUATORIAL SECTIONS OF THE
SURFACE OF ADVANCING STREAM

EARTH

DIRECTION OF ADVANCE
OF STREAM SURFACE

TO THE SUN

Figure 6.7. The Chapman and Ferraro model to explain solar storms, taken from Chapman and Ferraro,[27] 1931. Solar particles strike the earth's magnetic field and a cavity is formed about the earth. The cavity is not a permanent feature of the magnetic field.

the charged particles diffuse out, the earth's magnetic field returns to normal. A sketch of the Chapman–Ferraro model of the ring currents appears in Fig. 6.7.

A number of indicators of the intensities of magnetic storms have been derived from ground measurements. The variation of the earth's magnetic field is measured by K_p, the planetary three hour range index. It is the mean obtained from all available stations on a scale from 0 to 9.[28] D_{st} is the storm time disturbance magnetic field. It is the difference between the whole field at any instant and the field averaged over a month or more. It is a measure of the ring currents. The magnetic indices are summarized and published periodically.[29] Many interesting correlations have been obtained that assist in an understanding of magnetic storms and their effects on the earth's magnetic

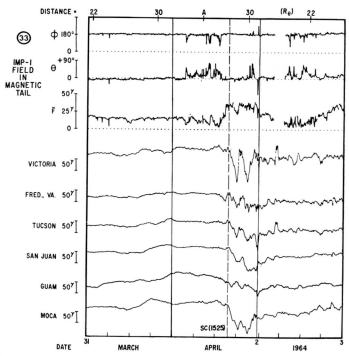

Figure 6.8. Correlations between magnetic field measurements on Imp 1 and disturbance magnetic fields measured at magnetic observatories during the magnetic storm on April 1, 1964, taken from Behannon and Ness,[30] 1966. The magnitude of the magnetic field $\bar{F}$, and the angles in ecliptic coordinates are given. The magnetic field in the tail increases in the main phase of the storm while the magnetic field at the earth decreases.

cavity. The tail magnetic field magnitudes and directions have been correlated with the local disturbance fields at ground stations. One such example[30] is given in Fig. 6.8, where the ecliptic magnetic field angles, ϕ and θ, and the magnetic field $\bar{F}$, measured on the Imp 1 satellite, are plotted versus time. The magnetic field, measured at six ground stations during the April 1, 1964 magnetic storm is also plotted. The magnetic field in the tail increased by about 25 γ during the main phase of the storm when the ground fields decreased by about 100 γ. The ground fields increased to their initial values in a few days as the tail fields returned to normal.

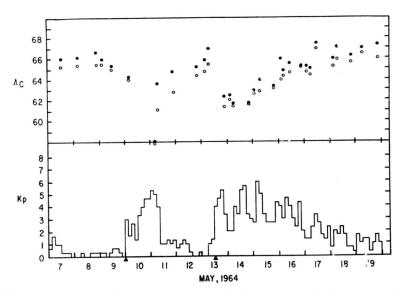

Figure 6.9. Correlations between the boundaries of trapped electrons and the magnetic field index, K_p, measured on the ground, taken from Williams and Ness, [31] 1966. At times of magnetic storms when K_p is large, the latitude of the upper boundaries of trapped particles decreases by a few degrees.

There is also evidence for a decrease by a few degrees of the northern latitude boundary of trapped radiation during magnetic storms. An example of the correlation between the trapped particle high latitude boundary and K_p versus time is given in Fig. 6.9.[31] The rather abrupt decrease in latitude is associated with a sharp rise in K_p. This could be the result of an increased plasma flow from the sun that causes more magnetic field lines to be concentrated in the magnetic cavity. These

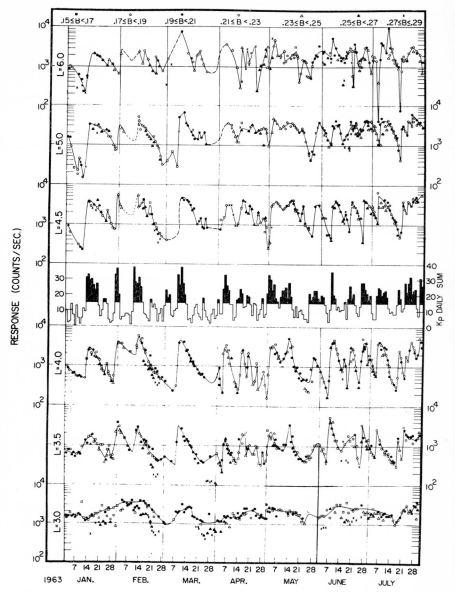

Figure 6.10. The flux of electrons with $E > 40$ keV measured on Injun 3 as a function of time, taken from Craven,[33] 1966. Data are given for $L = 3.0$ to 6.0 at the B values listed. The K_p daily sum is also plotted. Clear correlations are seen between K_p and the electron fluxes which rise and fall together.

additional lines compress the magnetic field. The trapped particles then move closer to the earth.

It was pointed out in 1962[32] that electrons with energies > 40 keV at $L > 3.0$ increase with K_p but that electrons with energies > 1.6 MeV decrease and then slowly increase and return to their original intensities. A comprehensive study of the variations of the low and high energy electron intensities with K_p is shown in Fig. 6.10 and 6.11.[33] The

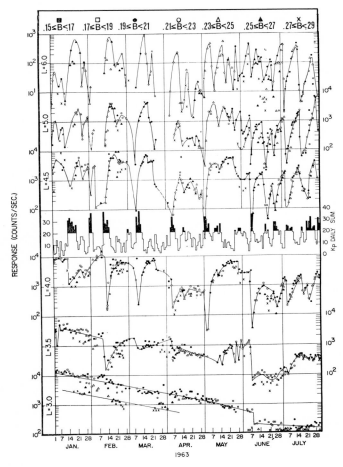

Figure 6.11. Fluxes of electrons with $E > 1.6$ MeV versus time measured on Injun 3, taken from Craven,[33] 1966. Data are given for $L = 3.0$ to 6.0 at the B values listed. The high energy electrons are anti-correlated with K_p. When K_p increases, the fluxes of 1.6 MeV electrons decrease, then gradually increase to their initial values.

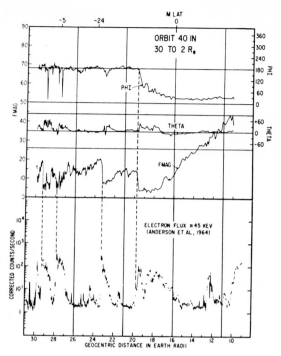

Figure 6.12. Correlations between electron fluxes and magnetic field measurements in the tail from Imp 1 on orbit 40 inbound, April 29, 1964, taken from Behannon and Ness,[30] 1966. The magnitude of the magnetic field $\bar{F}$, and the ecliptic angles θ and ϕ, are plotted. The neutral sheet was crossed at $19.8 R_e$ coincident with an abrupt increase in the electron flux.

correlation of the electrons with $E > 40$ keV, and the anti-correlation of electrons with $E > 1.6$ MeV is clear. The outer radiation belt of low energy electrons rises and falls in phase with the magnetic storms. The effect is small below $L = 3.0$ but is still present out to $L = 6.0$.

The pulses of electrons with $E > 40$ keV observed in the tail[19,20] on Imp 1 in April, 1964 have been correlated with the magnetic field measurements,[30] ϕ, θ, and $\bar{F}$ and are shown in Fig. 6.12. Here the tail field seems to be anti-correlated with the pulses of electrons. The magnitude of the magnetic field decreases as the bursts of electrons appear. At this time the spacecraft was close to the neutral sheet which it later crossed at $19.8 R_e$. The neutral sheet was probably the source of the electrons.

I. McDiarmid and J. Burrows[34] observed "spikes" of electrons with

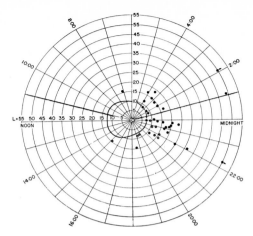

Figure 6.13. A polar plot of the positions of electron spikes measured at 1,000 km height versus local time and L, taken from McDiarmid and Burrows,[34] 1965. Note that the L values increase from the center of the diagram outward. The positions of the spikes are all at latitudes higher than the boundary of the outer radiation zone indicated by the dark line.

$E > 40$ keV at altitudes of 1,000 km near the midnight longitude. These were found on magnetic field lines that were connected to the tail of the magnetic cavity. The time distribution of these short pulses is shown on the polar plot of Fig. 6.13.[34] The location of each of the pulses is given by a black dot. Note that the L values run *outward* from the center of the polar plot, i.e., the *smallest* L values are on the inside. A smooth black curve is drawn at the high latitude boundary of the outer radiation zone for a height of 1,000 km. The spikes all occur at higher latitudes (higher L values) than the trapped radiation. They are definitely concentrated on the night side. Seventy per cent of the events have widths > 2 deg. Some of the pulses are as high as 10^9 electrons/cm²-sec-ster and 30 per cent have intensities $> 10^7$ electrons/cm²-sec-ster. The high latitude pulses occur at times of moderately high K_p and at times when electron fluxes in the outer radiation belt are higher than normal. There does not, however, appear to be a strong correlation between the magnitude of K_p and the magnitude of the electron pulses. The data are consistent with the particles originating in the vicinity of the neutral sheet.

The electron and proton fluxes in the tail of the magnetic cavity were also studied[35] during October 1962 with the ion-electron detector on Explorer 14. The maximum fluxes of protons with $E > 125$ keV, 2×10^4

protons/cm²-sec-ster, were found associated with simultaneous electron fluxes. The maximum intensities of electrons with $E > 20$ keV were about 8×10^6 electrons/cm²-sec-ster. The pulses showed a similarity with the electrons and protons observed in the aurora. A. Konradi[35] suggests that the similarity of the tail electron and proton energy distributions, and the similarity of the tail particles to the electrons and protons in the aurora, make it likely that the tail and aurora particles have a common source.

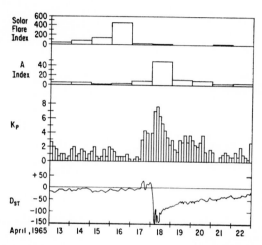

Figure 6.14. A plot of the solar flare index and the magnetic field indices A, K_p and D_{st} during the magnetic storm of April 17, 1965, taken from Cahill,[36] 1966. When the magnetic disturbances on the ground are large, as measured by K_p, the disturbance field is large and negative, indicating the presence of an inflated magnetosphere and a non-symmetric ring current.

There is strong evidence for the inflation of the inner magnetosphere during a magnetic storm. L. Cahill[36] observed the magnetic field on Explorer 26 during the great magnetic storm of April 17, 1965. The solar flare index and the magnetic field indices for that flare are given in Fig. 6.14. The magnetic storm followed the flare by the usual two days. K_p rose to 8 and D_{st} fell to $-150\,\gamma$. The magnetic field at Explorer 26 dipped to a minimum of $217\,\gamma$ below the normal field at $3R_e$. The greatest depression was in the evening quadrant. Magnetic field measurements in the critical region between 2 and $5R_e$ explained the magnetic storm observations on the ground. The sudden commencement was a compression of the whole magnetosphere during the initial phase of the storm.

The surface field dropped abruptly at 1730 UT, April 17, when the magnetosphere expanded. The main phase decrease started at 0200 UT at the minimum of the magnetic field. The satellite apparently passed through the heart of a highly inflated region of the magnetosphere during the minimum of the magnetic field. No evidence for a symmetrical ring current was found either from the ground magneto-grams or the outbound satellite measurements. The main phase of this storm was apparently caused by a large number of charged particles injected, or locally accelerated, in the evening sector of the magnetic cavity.

Since ground magnetic field measurements indicated that the storm drifted west and since we know that protons drift west around the earth in the earth's magnetic field, we conclude that low energy protons must have been principally responsible for the main phase. Protons with $E < 60$ keV satisfy the drift time requirements. The protons that cause the morning–evening asymmetry between 2 and $5R_e$ apparently never drifted completely around the earth. The rapid recovery back to normal indicated that the particles injected before 0800 UT were rapidly lost after 1000 UT. In the slow recovery phase, as the magnetic field returned to normal, there was a temporary axially symmetric ring current that decayed away with a time constant of four days.

It is interesting that there is still little direct experimental evidence for a permanent ring current. The magnetic field measurements have indicated primarily where the ring is not to be found.[36] Likewise, a search for a permanent ring of charged particles has been unsuccessful. The known trapped radiation belts are too low in flux to give the large changes in magnetic field that are measured at the earth.

6.5 THE AURORA

The auroral sheets of pale green light with occasional patches of red and pink are familiar to residents of the polar regions. This panorama of color changes from minute to minute and hour to hour throughout the long polar night. The green oxygen radiation at 5577 A and the violet radiation at 3900 A originates at an altitude of about 110 km. The red oxygen radiation at 6300 A originates primarily between 200 km and 400 km. Since the discovery of electrons and protons, this shimmering light has often been attributed to beams of particles coming from the sun. It was thought that particles entered the earth's magnetic field by

Figure 6.15. Photographs of the aurora taken from Akasofu,[37] 1965. In the upper left is a photograph of a quiet and diffuse auroral arc. At the top right is a similar arc with pink light from excited nitrogen molecules. The other four photographs are ribbon-like auroral types called active rayed bands.

spiraling down along interplanetary magnetic field lines that connected to the auroral zone at geomagnetic latitudes of 65 to 70 deg.

Photographs of the aurora taken from Akasofu,[37] 1965 are shown in Fig. 6.15.

The auroral forms occur in a ring centered about the geomagnetic poles with the greatest intensity at midnight. Two such aurora are shown in Fig. 6.16.[37] The first is a quiet homogeneous arc that is observed in normal quiet aurora. The second is the kind of display that occurs during the peak of a magnetic storm.

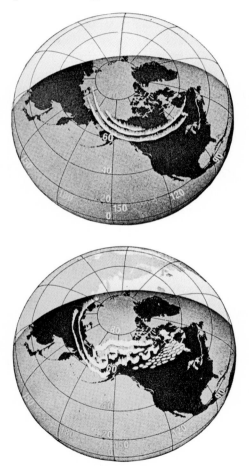

Figure 6.16. Quiet homogeneous arcs of the aurora before the magnetic storm, top, and after the magnetic storm, bottom. These storms originate in the midnight sector and usually last for two or three hours, taken from Akasofu,[37] 1965.

Most of the lines in the aurora have been identified. Of particular interest are the oxygen green 5577 A and the red 6300 A lines. An energy level diagram for the two excited states emitting these radiations is shown in Fig. 6.17. The transitions for both lines are forbidden so that the lifetimes are long—0.74 sec for the 5577 A green line and 110 sec for the 6300 A red line. At the lowest altitudes, below 200 km,

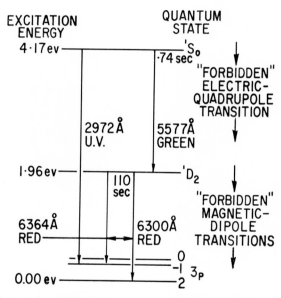

Figure 6.17. Energy level diagram of the two excited states of atomic oxygen. The green line 5577 A has a lifetime of 0.74 sec and the red line of 6300 A, 110 sec. If the molecular concentration is sufficiently high, atoms can be de-excited by molecular collision. Therefore, the red line is seen only at altitudes above 200 km. Taken from O'Brien,[42] 1966.

the oxygen atom collides with and loses energy to other molecules in a time short compared to 110 sec. Therefore, the red line is not seen and the green predominates. Above 200 km the atmospheric density is sufficiently low that collisions are infrequent. The excited state then decays to give the red line which is strong at the high altitudes.

An excellent summary of the positions of the auroral forms as a function of time is given by T. Davis.[38] During the IGY (International

Geophysical Year, 1957–58), all sky cameras obtained detailed information about the time and position variations of the auroral forms.   The
locations of five cameras utilized for statistical analyses located in
Alaska at 61.5, 64.5, 65.5, 66.5, and 68.5 deg geomagnetic latitude are
shown in Fig. 6.18.[38]   Positions that correspond to heights of 100 km

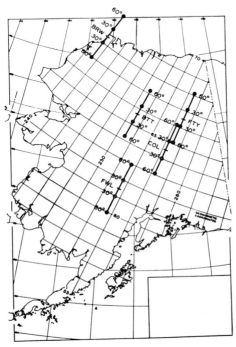

Figure 6.18. Locations of the five cameras used for auroral studies, taken from
Davis,[38] 1962.   Positions that correspond to heights of 100 km above the surface
of the earth at zenith angles of 0, 30, and 60 deg for each camera are shown.
The locations are Barrow (BRW), Bettles (BTT), Fort Yukon (FTY), College
(COL), and Farewell (FWL).

above the surface of the earth at zenith angles of 0, 30, and 60 deg for
each camera are shown.   The cameras and their meridian markers give
observations all along the geomagnetic meridian from 60 to 70 deg
north geomagnetic latitude.   The incidences of the auroral forms versus
geomagnetic latitude and local time are given in Fig. 6.19[38] for 59 nights
during 1957–58.   The maximum incidence occurs near local midnight
and at a geomagnetic latitude of 66.5 deg over Alaska.   The angular
width of the auroral zone at one-half peak intensity is 5 deg.

K

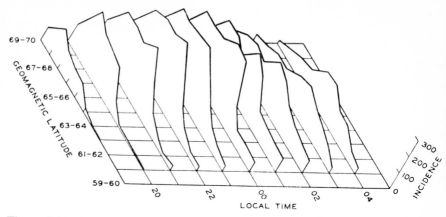

Figure 6.19. A plot of auroral incidences versus geomagnetic latitude and local time, taken from Davis, [38] 1962. Data from observing stations at 61.5, 64.5, 65.5, 66.5, and 68.5 deg geomagnetic latitude for 59 nights during 1957–58 were used.

The incidence of auroral forms is larger at times of high K_p. The auroral forms have net drifts along the meridian toward the south at geomagnetic latitudes of 60 to 70 deg. The alignment of the auroral formation is very striking, as is seen from a polar coordinate plot of geomagnetic co-latitude and geomagnetic time in Fig. 6.20.[39] The dashed parts of the lines represent the discontinuous morning post-break-up-aurora. During the evening hours, speeds of > 200 meters/ sec in a westward direction are seen. In addition, there is a clockwise motion as observed from above the North Pole. Near geomagnetic midnight apparently an abrupt reversal in motion occurs. This reversal toward the east usually happens first near 68 deg latitude and later at positions farther south. The visual auroral forms never overtake or cross one another. It is thought that the directions of the motions of aurora are intimately related to the directions of ionospheric currents. Fig. 6.20 then also indicates the directions of these ionospheric currents.

Direct measurements have been made of the particles producing visible auroras by Carl McIlwain[40] and by Richard Albert.[41] McIlwain[40] flew proton and electron detectors on two rockets which were fired into the aurora. He found that the electron fluxes were higher than the proton fluxes by more than a factor of 10. In one

AURORAL DISPLAYS OF 1957–1958, 2

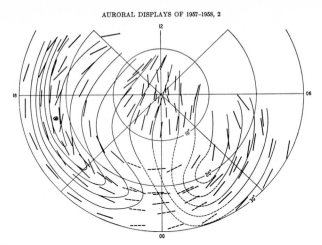

Figure 6.20. The alignment of auroral forms with geomagnetic co-latitude and geomagnetic time as polar and azimuthal angular coordinates, taken from Davis,[39] 1962. The average alignment, drawn with smooth curves, was deduced from the measured distribution of alignment lines. The dashed part of the lines represents the discontinuous post break-up aurora.

case he found that the electrons were concentrated in energy near 6 keV and in another case an exponential energy distribution was obtained from absorption in the atmosphere.

Previous measurements using the Doppler shift of the light from incoming protons as they captured electrons to become atomic hydrogen atoms had indicated that the protons must be near 10 keV. When viewed against the direction of the incoming protons the H line was shifted to higher frequencies. When viewed at 90 deg the line was broadened.

Albert [41] measured the energy distribution of electrons with an electrostatic spectrometer. An energy distribution is shown in Fig. 6.21 that is sharply peaked at 10 keV. It was obtained on a rocket flight from Fort Churchill at 250 km above a visible aurora on September 16, 1966. The count rate was averaged over the magnetic field pitch angles between 80 and 90 deg. The half width at half height is only 2.5 keV. Positive ion fluxes were several orders of magnitude lower in intensity and were not monoenergetic in energy.

Aurora also occur during the day. Radio waves reflected from the increased ionization monitor the daytime aurora. Aurora interrupt radio transmission in the amateur radio bands of 10's to 100's of

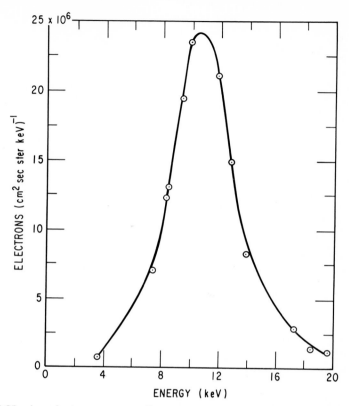

Figure 6.21. An electron energy distribution measured with an electrostatic spectrometer on a rocket in an aurora, taken from Albert,[41] 1966. The measurement was at 250 km over Fort Churchill, Canada on September 16, 1966.

megacycle/sec. Aurora also permit the reception of non-straight line television signals which otherwise go unaffected into space. A good review of the aurora and its effects on many physical phenomena is given by O'Brien.[42]

An understanding of the aurora requires an explanation of the source of the electrons, how they are accelerated to 10 keV and how they are guided to the auroral zone.

6.6 THEORY

The great number of measurements of the earth's magnetic field from ground and from satellite experiments are most useful when incorporated into a model of the earth's magnetic field. Such a model is valuable for computations and for collecting, correlating, and summarizing data. One useful model is the 1964 three dimensional model of George Mead.[43] He uses the magnetic field of a simple dipole collinear with the earth's axis of rotation and an external field due to surface currents on the boundary of the magnetosphere. A spherical harmonic expansion is used to obtain the field. The two most important coefficients are the one which gives a constant field parallel to the dipole and one which gives a constant field gradient along the earth–sun line. As expected, the field lines are compressed on the sunlit side of the earth and to a lesser degree on the dark side. At a critical latitude of 83 deg on the noon–midnight meridian, all field lines leaving the surface of the earth are bent back over the North Pole and cross the equator near midnight. They return in a symmetric manner over the South Pole.

The earliest theory to explain solar storms by particles from the sun was proposed by Sydney Chapman and V. Ferraro,[27] as shown in Fig. 6.7. They visualized the incident stream of charged particles near the earth's magnetic field as the approach of a good electrical conductor. Electrical currents are induced that shield the interior of the conductor from changes in the external magnetic field. The earth's magnetic field lines are pushed forward by the stream and the field is compressed. This increase in the magnetic field at the surface of the earth is called the "sudden commencement" of the magnetic storm. Since the stream of particles on the sunward side is retarded by the magnetic field a surface is formed. It bends around the space occupied by the earth's magnetic field to form a "hollow" or magnetic cavity. A ring current of charged particles is later formed around the earth which reduces the magnetic field at the surface of the earth during the "main phase" of the storm. The ring current dissipates in a few days and the magnetic field at the surface of the earth then returns to normal.

The modern theories began with J. Piddington[44] in 1960. In his view, during a magnetic storm the solar gas compresses the magnetic field on the sunlit side and sweeps the magnetic field lines near the poles back into a long tail, as indicated in Fig. 6.22. In his theory the lines in the tail are pulled out by the receding gas cloud and the tail magnetic field is reduced. That is, line L_3 shown dashed in Fig. 6.22, under normal conditions moves out to the solid line L'_3, line L_2 to L'_2, and L_1 to L'_1. He used this reduction to explain the decreases in the magnetic

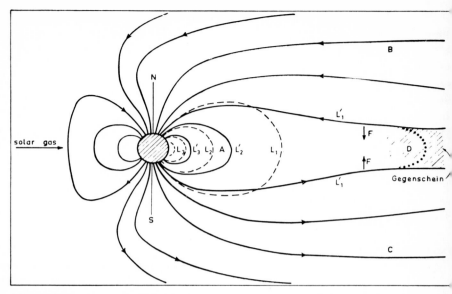

Figure 6.22. A magnetic cavity model to explain magnetic storms, taken from Piddington,[44] 1960. The lines of force at high latitudes during magnetic storms are drawn out to form a magnetic tail away from the sun. A line, normally at L_3, moves to L'_3, L_2 to L'_2, and L_1 to L'_1. This results in a decreased field in the tail. The tail is a temporary feature during the magnetic storm.

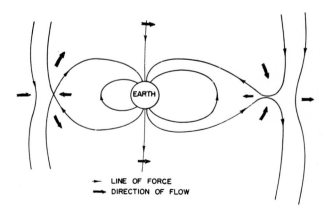

Figure 6.23. A fast magnetic merging model, taken from Dungey,[45] 1961. Merging of the interplanetary magnetic field with the earth's magnetic field occurs at the neutral point on the earth–sun line on the sunlit side of the earth. Lines of force are dragged across the North and South Poles and merge in the tail. The heavy arrows indicate the directions of movement of the magnetic field lines.

field at the earth during the main phase of the magnetic storm. During the recovery back to normal, the magnetic field lines reconnect and the magnetic field returns to a spherical shape. J. Piddington did not consider a permanent deformation of the magnetic cavity by a continuous solar wind.

The novel idea of annihilating magnetic field lines, called merging, was suggested by J. Dungey[45] as a way of introducing energy into the earth's magnetic cavity. His model of the magnetic field is given in Fig. 6.23. The solar wind with its south directed magnetic field approaches the earth from the left. The interplanetary magnetic field lines on the sunlit side of the earth merge with the earth's magnetic field lines at the neutral point on the sun–earth line. The magnetic field lines are swept to the rear of the earth by the solar wind and a long tail is formed. The lines coming over the North Pole merge with the lines coming over the South Pole in the magnetic equatorial plane.

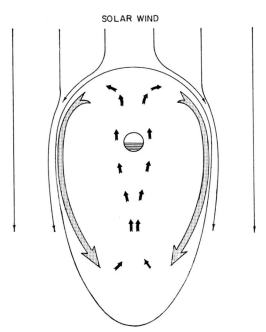

SOLAR WIND

Figure 6.24. Motions of magnetic field lines in the magnetic cavity caused by the viscous interactions with the solar wind, taken from Axford and Hines,[46] 1961. The magnetic field lines are pulled into the tail in the direction of the long arrows. The return flow to the sunlit side takes place in the direction of the short arrows. There is an additional effect, not shown here, due to the rotation of the earth.

Open magnetic field lines move out to the right to join the interplanetary field, while closed lines move to the left toward the earth. The arrows on the diagram indicate the direction of the movement of the magnetic field lines.

In 1961, W. Axford and C. Hines[46] published a detailed description of a magnetic cavity model that was driven by viscous tangential stresses at the boundary. Momentum is transferred across the boundary from the solar wind to particles in the magnetic cavity. They realized that the solar wind caused a permanent magnetic field distortion with a permanent tail. Magnetic field lines are pulled around into the tail in the directions indicated by the large, long arrows in Fig. 6.24. The return flow of the magnetic field lines to the sunlit side in the interior of the magnetic cavity follows the directions of the small arrows. An additional distortion of the magnetic field lines because of the rotation of the earth is superimposed onto the distortion created by the solar wind.

A significant contribution to the theory of merging was added by R. Levy, H. Petschek, and George Siscoe.[47] Their picture of combining and cutting of magnetic field lines is given in Fig. 6.25. Southward

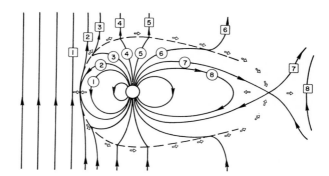

Figure 6.25. Fast merging of magnetic field lines, taken from Levy, Petschek, and Siscoe,[47] 1964. In this figure, the North Pole is down and the South Pole is up. The south-directed interplanetary magnetic field lines merge with the north-directed earth's magnetic field lines on the sun–earth axis on the sunlit side of the earth at the point indicated by the two facing hollow arrows. The magnetic field lines split, one-half moves north and one-half moves south. The hollow arrows show the directions of movement of the magnetic field lines from 1 to 2 to 3. . . . The magnetic field lines from the north and the south merge in the tail at the magnetic equator. Open field lines move to the right into interplanetary space. Closed magnetic field lines move to the left, rotate around the earth, and on to position 1 where the process repeats.

pointing magnetic field lines in the solar wind merge with north pointing magnetic field lines of the earth at the neutral point on the sunlit side. The interplanetary magnetic field lines are split in two; one-half progresses northward and one-half southward. The configurations of the lines at later times are identified by the numbers 1, 2, 3, 4, 5, 6, 7, and 8. As before, the lines coming over the North and South Poles are merged behind the earth. Open lines continue into interplanetary space and closed lines move toward the earth. The closed lines also rotate with the earth. After position 8, the lines move around the earth to the sun side where they progress out to the neutral point, join with the interplanetary field and the cycle continues again.

The above ideas were combined and extended by W. Axford, H. Petschek, and G. Siscoe[48] with their more detailed theory of merging. (See Fig. 6.26.) Again the tangential stresses move the magnetic field lines to the rear. Merging occurs at the nose of the magnetic cavity

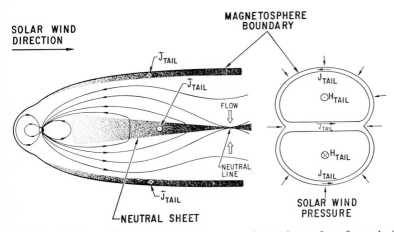

Figure 6.26. A fast merging model of the magnetic cavity, taken from Axford, Petschek, and Siscoe,[48] 1965. The interplanetary magnetic field lines merge at the nose of the magnetic cavity and are swept backward into the tail. Magnetic field lines from the north and the south merge on a neutral line perpendicular to the page. The hollow arrows show the direction of flow of the magnetic field lines toward merging. A neutral sheet or wedge is formed between the magnetic field lines that are directed toward the earth in the north and away from the earth in the south. An electric current in the neutral sheet directed out of the paper returns in the magnetic cavity boundary to form current loops that support the magnetic fields. A section through the tail looking away from the earth shows the directions of the currents in the boundary and the directions of the magnetic fields. Arrows at the cavity surface indicate the directions of the solar wind pressure.

and in the tail in the equatorial plane on a line perpendicular to the page. The neutral sheet separates the lines in the northern part of the tail from the lines in the southern part. Currents due to charged particles flow out of the surface of the paper in the tail and return in the bottom and top surfaces of the cavity. The magnetic field lines in the northern part of the tail above the neutral sheet point toward the earth and in the southern part below the sheet away from the earth. The neutral sheet is really a wedge that is thin at the merging line and tapers to several earth radii near the trapping region at $8R_e$. Note that this model requires merging in front on the sunlit side, as with the models of Dungey,[45] and Levy, Petschek, and Siscoe.[47] Since the neutral sheet has finite thickness, there is a small component of magnetic field from south to north and the neutral sheet is not entirely neutral.

Order of magnitude calculations by these authors[47,48] lend credibility to the merging model. Roughly one-half the magnetic field lines, that leave the earth from the auroral zone and higher latitudes, go backward into the tail. The others go forward into the magnetic cavity nose. The total magnetic flux through a semi-circular area near the pole is

$$\Phi = B \times \frac{\pi R^2}{2} . \tag{6.2}$$

Using $B = 0.6$ gauss and a radius at magnetic latitude of 65 deg of 3×10^8 cm, we find

$$\Phi = 6 \times 10^{16} \text{ Maxwell's.} \tag{6.3}$$

These lines go into the upper half-cylinder of the tail. For the magnetic flux in the tail, using $B = 20\,\gamma$ and $R = 20R_e$, we find

$$\Phi = 4 \times 10^{16} \text{ Maxwell's} \tag{6.4}$$

in good agreement with the flux through the polar area.

An electric potential across the tail from dusk to dawn is set up by the annihilation of the magnetic field at the neutral line. From Maxwell's equations the induced emf, V, is

$$V \text{ (volts)} = -\frac{300}{c}\frac{d\Phi}{dt}\left(\frac{\text{Maxwell's}}{\text{sec}}\right)$$

$$= -\frac{300}{c} BLv_B, \tag{6.5}$$

where L is the length of the line across the tail on which the merging occurs. It is perpendicular to the paper and about $40R_e$ (2×10^5 km) long. c is the velocity of light. The velocity of the magnetic field lines as they approach the neutral line is about $0.1v_A$[47] where v_A is the Alfvén velocity given by

$$v_A \text{ (cm)} = \frac{B \text{ (gauss)}}{\sqrt{4\pi m_i n}}. \tag{6.6}$$

n is the number of ions/cm³ and m_i is the mass of an ion in gm. Taking n as 10/cm³, and B as 20 γ, one finds $V = 30$ keV. The flow of energy is given by magnetic field energy density just outside the neutral sheet times three for the three degrees of freedom multiplied by the Alfvén velocity.

$$E = \frac{3B^2}{8\pi} v_A = \frac{3B^3}{8\pi} \frac{1}{\sqrt{4\pi m_i n}} \tag{6.7}$$

$$E = 7 \times 10^{-2} \text{ erg/cm}^2\text{-sec.}$$

The total energy flux over an area $1R_e$ in height and the width of the tail is

$$E_T = 7 \times 10^{17} \text{ erg/sec.} \tag{6.8}$$

This energy input from merging could supply the total normal energy loss in the aurora which is estimated to be $1-4 \times 10^{17}$ erg/sec. The merging energy is about an order of magnitude less than the energy flux in the solar wind of 5×10^{-1} erg/cm²-sec and the total solar wind energy flux over the nose of the magnetic cavity of 1×10^{19} erg/sec.

Since the merging energy is proportional to B^3, from Eq. (6.7), a doubling of the magnetic field in the tail increases the merging energy flux to the earth by a factor of 10.

It is of interest to compare the merging energy to other energy sources. The light energy from the sun striking the earth is 1×10^6 erg/cm²-sec or 2×10^{25} erg/sec over the magnetic cavity nose. And the energy of rotation of the earth is still much higher, 2×10^{36} ergs.

Objections[27] were raised to the fast merging theories because the mechanism for dragging the magnetic field lines into the tail at high velocities is not defined. The particle density is so low in the solar wind that the particle–particle collision time is a few years. If the friction at the magnetic field boundary takes place by particle–particle interactions, the merging time must also be a few years. Clearly this is too long if the tail magnetic fields must react to magnetic storms that occur every few days or weeks.

Instead Alex Dessler[49] suggested a magnetic cavity with a very long tail, many sun–earth distances in length. In his model, the merging is very slow and the energy input takes place all along the tail. There is no connection to the field lines of interplanetary space except at the end of the tail. Brian O'Brien[50] modified this model by connecting the neutral sheet to the aurora as shown by the dark cross-hatched region in Fig. 6.27. He divided the magnetic and charged particles into

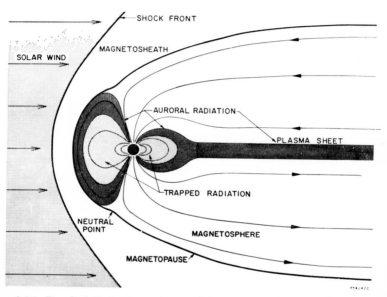

Figure 6.27. Dessler's[49] slow merging model of the magnetic cavity, as modified by and taken from O'Brien,[50] 1966. The neutral sheet is connected to the auroral region and is indicated by the dark cross-hatched area. The light cross-hatched area indicates the region of trapped particles. The earth is indicated by the black circle. The solar wind is incident from the left and forms the shock front. The magnetosheath is between the shock front and the magnetic cavity.

the following regions: The Van Allen geomagnetically trapped radiation region, the auroral region, the magnetosheath region, and the interplanetary space region. He emphasized the "auroral region", the entire region in the magnetic cavity except for the trapped radiation. The auroral particles, the islands or pulses in the tail, and the spikes at high latitudes at 1,000 km are all associated phenomena which occur in his "auroral region."

A detailed calculation of the motion of auroral particles in an electric and magnetic field model of the magnetic cavity has been carried out by H. Taylor and E. Hones.[51] The authors used their own magnetic field representation and an electric field deduced from the atmospheric current system that accompanies magnetic storms. They derived potentials as high as 50 keV that may be used to accelerate charged particles in the magnetic cavity. The surface of the magnetosphere is an equipotential surface since there is no connection with interplanetary lines of force. Their electric fields prohibit entry of solar wind protons into the tail but permit entry and acceleration of solar wind electrons on the evening side. They find a definite latitude boundary between auroral electrons and protons. The auroral protons enter the earth's atmosphere at the lower latitudes.

6.7 CONCLUSIONS

The primary observations that can be explained by a fast merging magnetic field model such as that of Axford, Petschek, and Siscoe[48] are now reviewed. The solar wind bombards the earth's magnetic field. The viscous forces on the boundary, unspecified, drag the magnetic field lines away from the boundary and create the long tail. The merging of magnetic field lines in the tail transforms the magnetic field energy into kinetic energy of electrons and protons. The closed magnetic field lines rotate with the earth to the sunlit side where they merge again with the interplanetary magnetic field lines. The continuous injection into the magnetic cavity of new magnetic field lines from interplanetary space with attached solar wind particles is a continuous source of replenishment of particles for the neutral wedge. Currents in the neutral wedge return through the boundary of the magnetic cavity to form current loops that furnish the magnetic fields in the tail.

The electron islands, spikes or pulses in the tail are electrons which are accelerated in the neutral sheet by the electric potential set up by the merging of the magnetic field lines. They are in transit toward or away from the trapped radiation belts. The intensity of these charged particles drops off with their distance away from the neutral sheet which indicates that the neutral sheet is the source of the particles.

The 10 keV electrons in the aurora obtain their energy from the tail potentials. The very sharp peaked energy distributions suggest that they originate from one location in the dusk to dawn neutral line.

Electrons are observed on the dawn side because the lines connecting to the dawn side of the tail are at a high negative potential while protons are observed on the dusk side because the lines connected to the dusk side are at a high positive potential. This is indicated on the sketch of the magnetic field line configuration of Fig. 6.28. In this model the

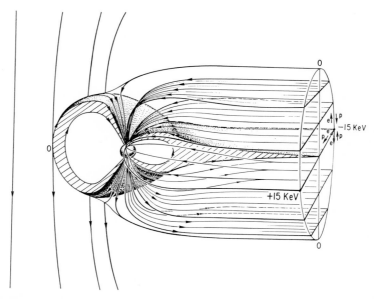

Figure 6.28. Magnetic field lines in a three-dimensional model of fast merging. The magnetic field lines merge on a line perpendicular to the page. The surfaces formed by the magnetic field lines can be seen at different times. The resultant electric potentials, arbitrarily set equal to 0 at the nose of the magnetosphere, are 0 on the noon–midnight meridian plane, $+15$ keV on the dusk side, and -15 keV on the dawn side of the merging line. The directions of the movements of electrons and protons in the neutral wedge and in the boundaries of the magnetic cavity are shown. The neutral wedge is connected to the auroral zone. A section through the auroral zones and the neutral wedge is indicated by the cross-hatched area.

potential is arbitrarily fixed at 0 in the nose of the magnetic cavity on the sun–earth line and is 0 on the meridian plane that passes through the sun–earth line. The potential of the tail at the line of merging is $+15$ keV on the dusk side and -15 keV on the dawn side. Aurora may also be observed in the day because electrons that are accelerated by the electric potential in the tail are carried around to the day side by the closed magnetic field lines which rotate with the earth. These particles also drift around the earth with a motion of their own.

The main phase of a magnetic storm is observed as a decrease in the magnetic field on the surface of the earth. It is caused by the inflation of the magnetosphere due to a large number of protons temporarily injected by a magnetic storm at L values from 2 to 5. These protons are injected in the dusk quadrant in a non-symmetric way.

A fuzzy boundary exists between the outer edge of the outer electron radiation belt and the auroral region. Magnetic field lines connect the auroral region to the neutral sheet. At times of magnetic storms more magnetic field lines are pulled into the tail. The magnetic field both fore and aft is compressed and the outer boundary of the outer electron radiation belt moves to lower latitudes. Most of the particles injected are immediately lost into the atmosphere and are not trapped except for short periods of time.

Some protons and electrons are trapped, however, and diffuse into the trapped radiation region. This is the source of the low energy protons trapped in the radiation belt. The protons are accelerated to 10 keV in the tail by the potential set up by the magnetic field merging. They then diffuse across the magnetic field lines by changing energy according to $EL^3 = $ constant while conserving their magnetic moments.

It was observed that the low energy electrons with $E > 40$ keV in the outer electron radiation belt increase and decrease along with K_p. During these times more new particles are injected into the magnetic cavity along with the greater number of magnetic field lines. The electrons are accelerated up to the energies of observation by the potential created during merging. However, the higher energy electrons with $E > 1.6$ MeV decrease as K_p increases and sometimes return to their initial values when the magnetic field returns to normal. This could be an adiabatic change in energy while conserving the magnetic moment. Sometimes it takes several weeks for the electron flux to return to normal. In this case, old electrons are probably lost and new ones diffuse inward from the magnetic cavity boundary and gain energy while conserving their magnetic moments.

Qualitatively most of the phenomena of the magnetic cavity, the aurora, magnetic storms and some of the phenomena of the radiation belts can be explained by the theory of fast merging of magnetic field lines. Order of magnitude energy calculations are reasonable. But there are many questions yet to be answered. First, there are no direct measurements of magnetic field line merging.   There are no measurements of electric fields that give the calculated accelerating potentials of 30 keV.   The mechanism of pulling the magnetic field lines into the tail is not explained.   No detailed calculations have been made to show what fraction of the electrons or protons in the solar wind cross the shock front, pass through the magnetosheath and enter the magnetic cavity.   No loss mechanisms have been combined with injection calculations to describe equilibrium particle space or energy distributions.   No time-dependent Fokker-Planck type equation has been solved to obtain the time distributions of the particles.   Until the crucial measurements are made and the detailed calculations are carried out, the theoretical explanations must remain incomplete.

REFERENCES

1. Chapman, S. and Bartels, J., *Geomagnetism*, Oxford Univ. Press, Oxford (1940).
2. Sonett, C. P., Judge, D. L. and Kelso, J. M., "Evidence concerning instabilities of the distant geomagnetic field: Pioneer I," *J. Geophys. Res.* **64**, 941–943 (1959).
3. Coleman, P. J., Jr., Davis, L., Jr., and Sonett, C. P., "Steady component of the interplanetary magnetic field: Pioneer V," *Phys. Rev. Letters* **5**, 43–46 (1960).
4. Ness, N. F., "Earth's magnetic field: a new look," *Science* **151**, 1041–1052 (1966).
5. Heppner, J. P., Ness, N. F., Scearce, C. S. and Skillman, T. L., "Explorer 10 magnetic field measurements," *J. Geophys. Res.* **68**, 1–46 (1963).
6. Cahill, L. J. and Amazeen, P. G., "The boundary of the geomagnetic field," *J. Geophys. Res.* **68**, 1835–1843 (1963).
7. Cahill, L. J., Preliminary results of magnetic field measurements in the tail of the geomagnetic cavity," *I. G. Bull.* **79**, *Trans. Am. Geophys. Union* **45**, 231–235 (1964).
8. Cahill, L. J., "The magnetosphere," *Scientific American*, March 1965, p. 58–68.
9. Ness, N. F., Scearce, C. S. and Seek, J. B., "Initial results of the Imp 1 magnetic field experiment," *J. Geophys. Res.* **69**, 3531–3569 (1964).
10. Ness, N. F., "The earth's magnetic tail," *J. Geophys. Res.* **70**, 2989–3005 (1965).
11. Biermann, L., "Kometenschweife und Solare Korpulscular Strahlung," *Zeit. F. Astrophys.* **29**, 274–286 (1951).
12. Parker, E. N., "Interaction of the solar wind with the geomagnetic field," *The Physics of Fluids* **1**, 171–187 (1958).
13. Gringauz, K. I., Bezrukikh, V. V., Ozerov, V. D. and Kybchinskii, R. E., "Study of the interplanetary ionized gas, high energy electrons, and solar corpuscular radiation by means of three electrode traps for charged particles on the second soviet cosmic rocket," *Soviet Physics Dokl.* **5**, 361–364 (1960).
14. Bonetti, A., Bridge, H. S., Lazarus, A. J., Rossi, B. and Scherb, F., "Explorer 10 plasma measurements," *J. Geophys. Res.* **68**, 4017–4063 (1963).

15. Snyder, C. W., Neugebauer, M. and Rao, V. R., "The solar wind velocity and its correlation with cosmic ray variations and with solar and geomagnetic activity," *J. Geophys. Res.* **68**, 6361–6370 (1963).
16. Frank, L. A., "A survey of electrons $E > 40$ keV beyond 5 earth radii with Explorer 14," *J. Geophys. Res.* **70**, 1593–1626 (1965).
17. Freeman, J. W., Jr., "The morphology of the electron distribution in the outer radiation zone and near the magnetospheric boundary as observed by Explorer 12," *J. Geophys. Res.* **69**, 1691–1723 (1964).
18. Gringauz, K. I., Kurt, V. G., Moroz, V. I. and Shklovskii, I. S., "Ionized gas and fast electrons in the vicinity of the earth and in interplanetary space," *Dokl. Akad. Nauk SSSR* **132** (5), 1062–1065 (1960).
19. Anderson, K. A., Harris, H. K. and Paoli, R. J., "Energetic electron fluxes in and beyond the earth's outer magnetosphere," *J. Geophys. Res.* **70**, 1039–1050 (1965).
20. Anderson, K. A., "Energetic fluxes in the tail of the geomagnetic field," *J. Geophys. Res.* **70**, 4741–4763 (1965).
21. Murayama, T., "Spatial distribution of energetic electrons in the geomagnetic tail," *J. Geophys. Res.* **71**, 5547–5557 (1966).
22. Montgomery, M. D., Singer, S., Conner, J. P. and Stogsdill, E. E., "Spatial distribution, energy spectra, and time variations of energetic electrons ($E > 50$ keV) at 17.7 earth radii," *Phys. Rev. Letters* **14**, 209–213 (1965).
23. Bame, S. J., Ashbridge, J. R., Felthauser, H. E., Olson, R. A. and Strong, I. B., "Electrons in the plasma sheet of the earth's magnetic tail," *Phys. Rev. Letters* **16**, 138–142 (1966).
24. Fitzgerald, G. F., "Sunspots and magnetic storms," *The Electrician* **30**, 48 (1892); "Sunspots, magnetic storms, comets' tails, atmospheric electricity, and aurora," *The Electrician* **46**, 287–288 (1900).
25. Lodge, O., "Sunspots, magnetic storms, comets' tails, atmospheric electricity, and auorora," *The Electrician* **46**, 249–250 (1900).
26. Dessler, A. J. and Michel, F. C., "Magnetospheric models," in *Radiation Trapped in the Earth's Magnetic Field*, p. 447–456, (ed.), B. M. McCormac, Reidel Publishing Co., Dordrecht-Holland (1966).
27. Chapman, S. and Ferraro, V. C. A., "A new theory of magnetic storms, Part I—the initial phase, and Part II—the main phase," *Terrest. Magnetism and Atmospheric Electricity* **36**, 77–97, 171–186 (1931); **37**, 147–156, 421–429 (1932), and **38**, 79–96 (1933).
28. Details concerning K_p may be obtained from Bulletin No. 12d, International Union of Geodesy and Geophysics, (ed.), J. Bartels, Washington (1950); Bartels, J., "Tagliche Erd Magnetische Charakterzahlen," *Abh. der Akad. d. Wiss in Göttingen*, Sonderheft, 1884–1950 (1951).
29. Almost every month geomagnetic and solar data is summarized in the Journal of Geophysical Research by J. Virginia Lincoln. These data can also be obtained from the World Data Center (Upper Atmosphere Geophysics) ESSA, Boulder, Colorado.
30. Behannon, K. W. and Ness, N. F., "Magnetic storms in the earth's magnetic tail," *J. Geophys. Res.* **71**, 2327–2351 (1966).
31. Williams, D. J. and Ness, N. F., "Simultaneous trapped electron and magnetic tail field observations," *J. Geophys. Res.* **71**, 5117–5128 (1966).
32. Pizzella, G., McIlwain, C. E. and Van Allen, J. A., "Time variations of intensities in the earth's inner radiation zone, October 1959 through December 1960," *J. Geophys. Res.* **67**, 1235–1253 (1962).

33. Craven, J. D., "Temporal variations of electron intensities at low altitudes in the outer radiation zone as observed with satellite Injun 3," *J. Geophys. Res.* **71**, 5643–5663 (1966).
34. McDiarmid, I. B. and Burrows, J. R., "Electron fluxes at 1,000 kilometers associated with the tail of the magnetosphere," *J. Geophys. Res.* **70**, 3031–3044 (1965).
35. Konradi, A., "Electron and proton fluxes in the tail of the magnetosphere," *J. Geophys. Res.* **71**, 2317–2325 (1966).
36. Cahill, L. J., "Inflation of the inner magnetosphere during a magnetic storm," *J. Geophys. Res.* **71**, 4505–4519 (1966).
37. Akasofu, S. I., "The aurora," *Scientific American*, December 1965, 54–62.
38. Davis, T. N., "The morphology of the auroral displays of 1957–1958, 1. Statistical analyses of Alaska data," *J. Geophys. Res.* **67**, 59–74 (1962).
39. Davis, T. N., "The morphology of the auroral displays of 1957–1958, 2. Detailed analyses of Alaska data and analyses of high latitude data," *J. Geophys. Res.* **67**, 75–110 (1962).
40. McIlwain, C. E., "Direct measurement of particles producing visible auroras," *J. Geophys. Res.* **65**, 2727–2747 (1960).
41. Albert, R. D., "Nearly monoenergetic electron fluxes detected during a visible aurora," *Phys. Rev. Lettsrs* **18**, 369–372 (1967).
42. O'Brien, B. J., "Auroral phenomena," *Science* **148**, 449–460 (1965).
43. Mead, G. D., "Deformation of the geomagnetic field by the solar wind," *J. Geophys. Res.* **69**, 1181–1195 (1964).
44. Piddington, J. H., "Geomagnetic storm theory," *J. Geophys. Res.* **65**, 93–106 (1960).
45. Dungey, J. W., "Interplanetary magnetic field and the auroral zone," *Phys. Rev. Letters* **6**, 47–48 (1961).
46. Axford, W. I. and Hines, C. O., "A unifying theory of high latitude geophysical phenomena and geomagnetic storms," *Canadian J. Physics* **39**, 1433–1464 (1961).
47. Levy, R. H., Petschek, H. E. and Siscoe, G. L., "Aerodynamic aspects of the magnetospheric flow," *American Inst. Aeronaut. Astronaut J.* **2**, 2065–2076 (1964).
48. Axford, W. I., Petschek, H. E. and Siscoe, G. L., "Tail of the magnetosphere," *J. Geophys. Res.* **70**, 1231–1236 (1965).
49. Dessler, A. J., "Length of magnetospheric tail," *J. Geophys. Res.* **69**, 3913–3918 (1964).
50. O'Brien, B. J., "Interrelations of energetic charged particles in the magnetosphere," *Proceedings of Inter-Union Symposium on Solar-Terrestrial Physics* held in Belgrade, Yugoslavia, 29 August to 2 September 1966.
51. Taylor, H. E. and Hones, E. W., Jr., "Adiabatic motion of auroral particles in a model of the electric and magnetic fields surrounding the earth," *J. Geophys. Res.* **70**, 3605–3628 (1965).

Definition of Symbols

A	Angström. $1\,\text{A} = 10^{-8}$ cm.
A	Atomic weight.
A	Area.
A_0	Avagadro's number.
AU	Astronomical unit.
a_d	Drag acceleration.
a_p	Geomagnetic disturbance index.
B	Magnetic field.
B_b	Magnetic field at position b.
B_i	Interstellar magnetic field.
B_0	Average magnetic field.
$B_\parallel$	Parallel magnetic field.
$B_\perp$	Perpendicular magnetic field.
BeV	Billion electron volts, energy.
BV	Billion volts, rigidity.
C	Constant.
C_d	Drag coefficient.
CRAND	Cosmic Ray Albedo Neutron Decay.
c	Velocity of light.
D	Dispersion of a whistler.
D_{st}	Daily magnetic field variation on the earth during magnetic storms.
sub $_d$	Daughter.
$\mathscr{E}, \mathscr{E}$	Electric field.
E	Energy.
$E_\perp$	Kinetic energy associated with the velocity perpendicular to the magnetic field.
E_M	Magnetic energy.

e	Electron charge. $e = 4.8 \times 10^{-10}$ esu.
eV	Electron volt, energy.
dE/dx	Energy loss/unit path length.
$\mathbf{F}$	Force.
F_d	Drag force on a sphere.
$\bar{F}$	Magnitude of the magnetic field.
$\mathcal{F}$	Constant.
f	Frequency of electromagnetic wave.
f	Number of individual sunspots.
f_{Be}	Electron cyclotron frequency.
f_{Bp}	Proton cyclotron frequency.
f_c	Cut-off frequency for an electron whistler.
f_{F2}	$F2$ layer cut-off frequency.
f_0, f_N	Ionosphere radio cut-off frequency.
$f(v)$	Function of velocity.
G	Gravitational constant. $G = 6.668 \times 10^{-8}$ dynes-cm^2/gm^2.
g	Acceleration of gravity.
g	Number of sunspot groups.
sub$_{\mathrm{H}}$	Hydrogen.
sub$_H$	Horizontal.
HF	High frequency radio waves.
h	Height.
h_p	Perigee height.
h_∞	Height at top of the atmosphere.
I	Radiation intensity.
i	Angle of incidence.
i	Angle of inclination of the satellite orbit to the equator.
J	Flux of solar protons.
$\mathbf{j}$	Current density, (amp/cm^2).
j_n	Albedo neutron flux at the position of injection.
j_p	Proton trapped flux.
K_p	Geomagnetic three hour range index.
$\sum K_p$	Sum of the 8 values of K_p as tabulated by Bartels in $J.$ *Geophys. Res.*
°K	Degree Kelvin.
k	Boltzman constant, $k = 1.38 \times 10^{-16}$ erg/deg.
k	Absorption coefficient.
k_{ob}	Efficiency assigned to each laboratory or observer.
keV	Thousand electron volts.

L	Magnetic L shell number. Distance to the magnetic shell at the equator in R_e.
L	Diameter of the magnetic cavity tail.
L_{sg}	Width of a supergranule.
LHR	Lower hybrid resonance.
$\mathscr{L}$	Effective modulation distance.
l	Characteristic dimension.
M	Mean molecular weight.
$M_\odot$	Mass of the sun.
MeV	Million electron volts, energy.
MJD	Modified Julian Days.
$\mathscr{M}$	Magnetic moment of the earth.
m	Mass.
m_e	Electron mass.
m_i	Mass of a molecule of identity i.
$\overline{m}_i$	Mean ionospheric effective mass.
m_p	Proton mass.
N_e	Number of electrons/cm³.
N_i	Number of ions/cm³.
N_j	Number of neutral atoms or molecules/cm³.
N_{sg}	Number of supergranulations formed at a position in time t.
N_T	Total number of electrons or ions/cm³.
NASA	National Aeronautics and Space Administration.
n	Number of atoms/cm³.
n_{cr}	Number of cosmic ray protons/cm³.
n_i	Number of molecules/cm³ of identity i.
n_0	Number of plasma ions/cm³ at zero thermal velocity.
n_s	Number of solar wind protons/cm³.
sub$_0$	Initial position.
P	Penumbra.
P	Power.
$P(f)$	Magnetic field power as a function of frequency.
$\mathscr{P}$	Period.
p	Pressure.
p.e.f.	Precipitated electron flux.
p_i	Interstellar pressure.
$p_{photosphere}$	Gas pressure in the photosphere.
$p_{sunspot}$	Gas pressure in the sunspot.
q	Electrons produced/cm³-sec.

q_i	Coefficient of thermal diffusion.
R	Wolf relative sunspot number.
R	Rigidity.
R_e	Radius of the earth.
R_s	Radius to the boundary of the solar wind cavity.
$R_\odot$	Radius of the sun.
r	Distance.
r	Angle of refraction.
r_B	Cyclotron radius.
$\mathbf{r}_n$	Distance at which solar wind becomes supersonic.
r_s	Solar wind distance.
r_{se}	Sun–earth distance.
r_{sh}	Shock distance.
S	Ratio of solar wind velocity to the thermal velocity parallel to the magnetic field.
SPAND	Solar Proton Albedo Neutron Decay.
S_D^-	Solar daily magnetic field variation on the earth during quiet times.
T	Temperature.
T_e	Electron temperature.
t	Time.
t_d	Diffusion time.
Δt	Time delay.
Δt_{sg}	Life time of a supergranule.
U	Umbra.
U	Internal thermal energy.
UT	Universal time.
UV	Ultraviolet radiation.
V	Volts.
V_c	Coulomb energy.
V_g	Gravitational energy.
VHF	Very high frequency radio waves.
v	velocity.
v	neutron or proton velocity.
v_A	Alfvén velocity.
v_{cr}	Cosmic ray velocity.
v_n	Sound velocity.
v_s	Solar wind velocity.
v_s	Satellite velocity.

$v\sigma$	Southward drift velocity.
v_t	Thermal velocity.
v_{up}	Upward drift velocity.
$v_\perp$	Velocity perpendicular to the magnetic field.
$v_{\parallel}$	Velocity parallel to the magnetic field.
y	Elapsed time in years.
Z, Z_1, Z_2	Charge of nuclei.
α	Pitch angle. Angle between the magnetic field and the particle directions.
α	Recombination coefficient (cm³/sec).
α_0	Pitch angle at the equator.
β	Ratio particle kinetic energy density to magnetic field energy density.
β	Velocity divided by the velocity of light.
γ	c_p/c_V, the ratio of the specific heat at constant pressure to the specific heat at constant volume.
γ	10^{-5} gauss.
γ	$1/(1-\beta^2)^{1/2}$
γ group	A multipolar magnetic spot group on the sun.
δ	A constant.
ϵ	Angle of scattering by magnetic field irregularities.
ζ	Angle between the interplanetary magnetic field and the direction to the sun.
η	Magnetic fluctuation diffusion variable.
η	Number of magnetic irregularity scatters.
θ	Angle of the magnetic field with the sun–earth line in the noon–midnight meridian.
K	Diffusion coefficient.
$K_{\parallel}$	Diffusion coefficient parallel to the magnetic field.
$K_\perp$	Diffusion coefficient perpendicular to the magnetic field.
λ	Latitude.
λ	Wavelength.
λ_{cr}	Cosmic ray mean free path.
$\lambda_\odot$	Latitude of the sun.
μ	Magnetic moment, first adiabatic invariant.
μ	Index of refraction.
v	Frequency of electromagnetic radiation.
sub_π	Parent.
ξ	Ratio of the thermal energy to the kinetic streaming energy.

ρ	mean atmospheric density, gm/cm^3.
σ	cross-section.
σ_i	electrical conductivity.
σ_{ij}	cross section of photons with neutral atoms or molecules.
σ_s	Stefan–Boltzman constant, $\sigma_s = 5.67 \times 10^{-5}$ erg/cm^2-sec-deg^4.
τ	Time between collisions.
τ	Neutron lifetime, $\tau = 1.0 \times 10^3$ sec.
Φ	Magnetic flux.
Φ_T	Direction of the axis of solar wind temperature diagrams.
ϕ	Angle of the magnetic field with the sun–earth line in the ecliptic plane of the earth.
ϕ	Longitude of the sun.
ϕ_0	Initial longitude of the sun.
χ	Injection coefficient.
χ	Angle of the magnetic field direction to the tangent to the earth.
ψ	Angle of inclination of the magnetic field.
Ω	Angular rotation velocity of the earth.
$\Omega_\odot$	Angular rotation velocity of the sun.
ω	Angular frequency, $2\pi f$.
ω_{12}	Cross-over angular frequency for the electron and proton whistlers.
ω_B	Cyclotron angular frequency.

Author Index

Subject Index

I 2